ANAEROBIC TREATMENT OF INDUSTRIAL WASTEWATERS

ANAEROBIC TREATMENT OF INDUSTRIAL WASTEWATERS

Edited by

Michael F. Torpy

Energy and Environmental Systems Division
Argonne National Laboratory
Argonne, Illinois

NOYES DATA CORPORATION
Park Ridge, New Jersey, U.S.A.

Library of Congress Catalog Card Number 88-2047
ISBN: 0-8155-1165-5
ISSN: 0090-516X
Printed in the United States

Published in the United States of America by
Noyes Data Corporation
Mill Road, Park Ridge, New Jersey 07656

10 9 8 7 6 5 4 3 2 1

Library of Congress Cataloging-in-Publication Data

Anaerobic treatment of industrial wastewaters / edited by Michael F.
Torpy (Energy and Environmental Systems Division, Argonne National
Laboratory, Argonne, Illinois).
p. cm. -- (Pollution technology review, ISSN 0090-516X ; no.
154)
Includes bibliographies and index.
ISBN 0-8155-1165-5
1. Sewage sludge digestion--Congresses. 2. Bacteria, Anaerobic-
-Congresses. I. Torpy, Michael F. II. Argonne National
Laboratory. Energy and Environmental Systems Division. III. Noyes
Data Corporation. IV. Series.
TD769.A537 1988 88-2047
628.3'51--dc19 CIP

Foreword

This book describes the latest advances in anaerobic treatment of industrial wastewaters. The technology of anaerobic digestion has developed significantly in recent years. It has gained increased credibility from successful applications in industry as well as other sectors.

Anaerobic biological waste treatment offers advantages over aerobic systems in terms of lower energy requirements, less biological sludge production, and the potential for energy recovery in the form of methane gas. The development of innovative reactor designs, based on the optimization of growth and retention of anaerobic microorganisms, has created an impetus to reevaluate the anaerobic treatability of many industrial waste streams.

The most recent and significant advances in anaerobic digestion are related to the technology's ability to accommodate relatively high rates of organic loading. Companies are also interested in using anaerobic digestion for the biodestruction of organic materials that are not removed in conventional aerobic treatment. As applications of anaerobic technology to various process streams increase, more successes are inevitable, resulting in industries that are more economically competitive because of their more judicious use of natural resources.

Companies adopting anaerobic digestion fall into the following categories: companies scheduled to expand their production-process capacity and whose treatment facilities are already at capacity; companies that discharge to publicly owned treatment works whose surcharge for treatment has increased substantially; companies that have relatively monotonous, highly concentrated organic waste streams contributing a major portion of total waste load; and companies in areas where extremely high land costs make conventional aerobic digestion too expensive.

In addition to companies that adopt anaerobic digestion as the treatment method of choice, several companies maintain active research programs to evaluate new methods, including anaerobic digestion, for treating new and existing waste streams. Two important reasons for such programs are (1) anticipated changes in environmental statutes and (2) the need to develop more cost-effective treatment techniques.

The information in the book is from *Anaerobic Treatment of Industrial Wastewaters,* edited by Michael F. Torpy of Argonne National Laboratory for the U.S. Department of Energy, issued August 1987. These are the Proceedings of the Second National Conference, held in Chicago, IL, September 1986.

The table of contents is organized in such a way as to serve as a subject index and provides easy access to the information contained in the book.

Advanced composition and production methods developed by Noyes Data Corporation are employed to bring this durably bound book to you in a minimum of time. Special techniques are used to close the gap between "manuscript" and "completed book." In order to keep the price of the book to a reasonable level, it has been partially reproduced by photo-offset directly from the original report and the cost saving passed on to the reader. Due to this method of publishing, certain portions of the book may be less legible than desired.

Preface

The technology of anaerobic digestion has developed significantly in recent years. It has gained increased credibility from successful applications in industry as well as other sectors. Although the technology is not without faults, it often represents an alternative to waste treatment that can provide substantial savings to industry. Many aspects of the technology contribute to such savings. One important aspect is the land cost saved by installing an anaerobic reactor instead of using an aerobic activated sludge process. Reduced sludge volumes, decreased energy requirements for aeration, and energy credit from the methane produced in the bioconversion, traditionally cited as sources of savings, are increasingly important considerations.

The most recent and significant advances in anaerobic digestion are related to the technology's ability to accommodate relatively high rates of organic loading. Companies are also interested in using anaerobic digestion for the biodestruction of organic materials that are not removed in conventional aerobic treatment. As applications of anaerobic technology to various process streams increase, more successes are inevitable, resulting in industries that are more economically competitive because of their more judicious use of natural resources.

Companies adopting anaerobic digestion seem to fall into the following categories:

- Companies scheduled to expand their production-process capacity and whose treatment facilities are already at capacity;
- Companies that discharge to publicly owned treatment works whose surcharge for treatment has increased substantially;
- Companies that have relatively monotonous, highly concentrated organic waste streams contributing a major portion of total waste load; and
- Companies in areas where extremely high land costs make conventional aerobic digestion too expensive.

In addition to companies that adopt anaerobic digestion as the treatment method of choice, several companies maintain active research programs to evaluate new methods, including anaerobic digestion, for treating new and existing waste streams. Two important reasons for such programs are (1) anticipated changes in environmental statutes, and (2) the need to develop more cost-effective treatment techniques.

The state of the art in anaerobic treatability testing is relatively simple. An anaerobic culture is mixed with the substrate in the presence of micronutrients. Samples to be tested must be maintained in an oxygen-free environment. Usually, the air of the testing vessel is replaced with a mixture of carbon dioxide and nitrogen gas and the vessel is sealed to prevent contamination by ambient air. Subsequently, the activity of the bioprocess is monitored by measuring the quantity and quality of biogas produced.

Like any biological process, anaerobic organisms may require an acclimation period (depending on the origin of the culture) before their metabolic activity can be measured. Conventional periods of 30 days may sometimes be too short, and potential applications may remain unrecognized because the testing period was not long enough. It is not unreasonable to measure the onset of biogas production after 100 days of testing, but periods of acclimation are usually much shorter.

Because the technology has been applied to a relatively limited array of waste types, caution should be exercised in drawing conclusions from results of a negative assay. The appropriate source of a culture often makes the difference between a successful assay and a failure to produce methane.

Lack of sophisticated water management within a facility is one of the largest obstacles to the adoption of anaerobic digestion. The more characteristics that are known about the waste stream, the better the chances for successful treatment. Segregation and testing of individual streams are commensurate with a successful testing program. Inorganic compounds such as cyanides, heavy metals, sulfates, and nitrates can be removed through pretreatment and render an otherwise untreatable sample biologically degradable.

The contributions in these proceedings show clearly that the technology of anaerobic digestion and its market are developing rapidly. Process engineering, reactor design and configuration, microbiology, and other technical concepts of anaerobic digestion are advancing through the world and in many facets of industry. The relative immaturity of the technology (compared to aerobic technologies), however, reminds us that innovation is required. Although the risks of adopting anaerobic digestion are often considered greater than those associated with conventional methods, the pay-off for installing the technology is usually substantial.

Some of the presentations in these proceedings disclose information never presented to the public before, and they indicate the favorable as well as unfavorable aspects of the technology. The information presented here is the result of personal and unselfish efforts by the authors. On behalf of the readers of the proceedings, I extend our sincere compliments to the presenters for their valuable and timely contributions.

Michael F. Torpy

NOTICE

Contents and Subject Index

ADVANCES IN ANAEROBIC BIOTECHNOLOGY FOR INDUSTRIAL WASTEWATER TREATMENT

R.E. Speece

Betz Chair Professor of Environmental Engineering
Drexel University
Philadelphia, Pennsylvania

ABSTRACT

Anaerobic biotechnology for industrial waste treatment is steadily expanding in the U.S. and abroad. Applications were initially in the area of food processing wastewaters. Meat packing plant wastewaters received most of the initial attention and rather simple anaerobic ponds were the most common type of unit process. The anaerobic contact process evolved from this effort. Subsequently, the key role of cell immobilization was recognized and the anaerobic upflow filter, upflow anaerobic sludge blanket (UASB) and fluidized bed unit processes evolved. Recently, hybrids of these first two processes have emerged to capitalize on the positive features of each. Volumetric loading rates of approximately 2 to 4 kg/m^3.d characteristic of the anaerobic contact process to 10 to 20 kg/m^3.d for the UASB. One short term loading rate of 125 kg/m^3.d was observed with a fluidized bed using a paper mill condensate wastewater. With the improvement in cell immobilization for unit processes for wastewaters comprised mainly of soluble organic pollutants, hydrolysis is often not rate limiting and conversion of volatile acids and hydrogen to methane becomes rate limiting. Therefore, further improvements in process efficiency may be linked to improvements in the specific activity of these classes of organisms. Laboratory studies using continuous cultures of methanogens converting acetate to methane have shown very significant stimulation resulted from supplementation with specific nutrients. The predominant population shifted from the slower growing, lower activity Methanothrix to the faster growing, higher specific activity Methanosarcina subsequent to the supplementation of iron, cobalt, nickel and trace levels of vitamin B_{12}. Generation times were reduced and acetate utilization rates were markedly increased. Acetates utilization rates of 35 kg/m^3.d were noted at solids retention times of as low as 5 days. Field studies are planned to evaluate the potential benefit with two phase digestion.

INTRODUCTION

It can now be comfortably said that anaerobic treatment is a well recognized industrial wastewater treatment technology. We have come through a rather prolonged period of skepticism where it was commonly considered that anaerobic treatment should be confined to the treatment of domestic wastewater sludges. Particularly in Europe, a large number of full scale anaerobic processes have been in operation for several years. These were mainly adapted to food processing wastewaters. The number of installations in the U.S. is also increasing. We are also seeing same bravado in the application of anaerobic treatment to rather difficult wastewaters. Suidan (1986) has reported the successful treatment of full strength coal conversion wastewater in an expanded bed granular activated carbon fixed film reactor at 21 day carbon and cell residence time. This is a significant milestone which demonstrates the synergism of carbon removal of toxic components to enhance the biological removal of the degradable components. In another case, anaerobic treatment was successfully applied to a textile wastewater which caused uncontrollable foaming in the aeration basin of an aerobic treatment plant, negating successful treatment. By proper adaptation, anaerobic treatment has been applied to organically polluted high salt concentration wastewaters. Another area of application is to use anaerobic treatment of wastewaters containing volatile organic contaminants which would tend to be air stripped in an aeration basin or trickling filter if aerobic treatment was used. A much reduced level of gas stripping occurs in anaerobic treatment and in addition, any stripped volatile organics can be combusted with the methane gas in a boiler or waste gas flare. A final example of successful adaptation of anaerobic treatment to biodegradation of an adiponitrile wastewater is quite notable.

At the present time, no clear "standard model" process configuration has emerged which is appropriate for all applications. The Upflow Anaerobic Sludge Blanklet (UASB) process appears to be most popular with the largest number of installations. However, in some cases e.g. pharmaceutical wastewaters, development of the "granules" was not achieved even after two years. In some cases, the granules have floated instead of settling. In many cases, the granules settle properly and provide a very effective means of cell immobilization. There are versions of the anaerobic contact process which are successfully operating, however, the biomass concentrations and associated organic loading rates are relatively low. Fixed film fluidized beds have demonstrated exceptionally high organic loading rates in some cases. Recycle is normally required to maintain fluidization and the density and size of media determines the required upflow velocity for fluidization. Sand of a size comparable to that used for water filtration requires relatively high upflow rates for fluidization e.g. over 10 gpm/ft^2. However, the tendency is to utilize smaller particle sizes and less dense particles to avoid excessive energy requirements for maintenance of fluidization. Granular carbon and volcanic pumice are favored. In Japan, a fluidized be of pumice constructed in a 50 feet deep reactor shaft excavated below grade is proposed in Japan with a required fluidization upflow velocity of one foot per minute giving a 50 minute hydraulic contact time requiring no recycle, suited for low strength wastewaters e.g. raw domestic wastewater.

It is clear that site specific conditions control the process configuration of choice. Type of wastewater, temperature, concentration of organic pollutant, toxic or refractory components, required degree of treatment, shock loading characteristics of hydraulic, organic or toxic nature, space available, segregation of wastewater flows and many other factors will dictate the most appropriate process configuration. Even though high rate anaerobic process configurations result in the lowest required reactor volume, it may be more economical to choose a low rate process that can be economically constructed in a membrane covered earthen basin. Low rate processes have inherently greater shock absorbing potential, requiring less supervision, control and often negate the need of flow or organic load equalization. In addition, there is a basic philosophical decision to be made on flow equalization. It may be more economical to incorporate the desired flow equalization volume in the biological reactor than to provide a comparable volume in a separate reactor. In some cases of organic acidic wastewaters, a separate equalization tank may require stainless steel tank construction, whereas ordinary carbon steel or even earth may be used for the biological reactor which incorporates the requisite equalization volume.

In this presentation, I wish to address two questions:

1) Where is the anaerobic process headed?
2) What further developments should be forthcoming?

I will dwell mainly on two aspects:

1) Cell immobilization and/or solids separation and retention.
2) Increases in the specific activity of the biomass.

CELL IMMOBILIZATION

The success of biological treatment is directly related to the efficiency of cell immobilization and/or retention. The cells must be retained in the system and higher cell concentrations result in lower reactor volume requirements. The original anaerobic processes utilized completely mixed reactors with no cell separation and recycle. Consequently the hydraulic retention times were equal to the solids retention times and the reactor volumes were necessarily very large. The development of the anaerobic contact process for meat packing wastewaters was a milestone because it incorporated cell separation and recycle, and uncoupled solids retention time from hydraulic retention time and reactor volume.

Young and McCarty's development of the attached growth upflow filter was another milestone, while Jeris' development of the fluidized bed biological system was still another milestone. Lettinga's development of the UASB significantly advanced the alternatives available for process configuration. The UASB was also an economical alternative to the relatively expensive packing requirement of upflow filters. As an aside, it appears to have been rather fortuitous that sugar beet wastewater was the candidate with which the UASB process was developed, because it provided the required hydrogen generation and calcium which appear to be requisites for granule formation (Lowenthal, Dold and Marais 1986).

Due to the inherently lower efficiencies and loading rates of the anaerobic filter and the occasional instability of the UASB, a hybrid of these two systems is being proposed- an unpacked lower section and a packed upper section. Such a hybrid avoids the potential plugging and higher costs of a fully packed reactor and yet maintains the high loading rates of a UASB while incorporating stability to the blanket by insuring against high solids loss from a filamentous blanket and complete loss of a granular blanket which for some reason floats instead of settles.

There is a fundamental deficiency in providing such an inefficient solids separation device as a crude gravity sedimentation chamber to such a low synthesis process as anaerobic treatment. A crude gravity sedimentation chamber in a high synthesis ratio aerobic process merely degrades the quality of the effluent. However, when synthesis rates are nominally less than 5%, as can occur in anaerobic processes, instead of the nominal 50% characteristic of aerobic processes, biomass inventory can be significantly depleted and start up and/or recovery times can be greatly prolonged. A hybrid reactor greatly decreases the dependence on development of a granular sludge for process stability, since the packing aids in retention of either flocculent or filamentous biomass should it predominate.

In my judgment, the greatest need in anaerobic treatment is to reduce start up times and increase process stability. It is unusual that a vendor can market a process which may require six months and even longer in some cases to achieve design loading rates, even more so, that a customer would tolerate such a deficiency. Enhanced biomass inventory build-up techniques should warrant the highest priority in development of anaerobic treatment.

To illustrate the crucial role of biomass retention efficiency, consider the following example of two scenarios:

1) Complete retention of the original seed and all subsequent synthesized biomass.
2) 20% retention of original seed and 20% retention of all synthesized biomass.

Assume in both cases that the design loading rate is 10 kg/m^3-day the specific activity is 1 kg COD/kg biomass-day, the bacterial synthesis rate is 5% and both reactors are started with 100% seed whose concentration of biomass is 3 kg/m^3. which is characteristic of digested domestic sludge.

In the first case, the time required to achieve design loading rate is:

$$3 \text{ kg/m}^3 \ (1 + 0.05)^n = 10 \text{ kg/m}^3$$
$$n = 25 \text{ days.}$$

In the second case which is somewhat characteristic of seeding an upflow filter or unpacked UASB, the required time to achieve the design loading rate is:

$$20\% \times 3 \ \text{kg/m}^3 \ (1 + .2 \times .05)^n = 10 \text{ kg/m}^3$$
$$n = 280 \text{ days.}$$

If the biomass retention rate could be improved to only 50%, the required start up time could he reduced to n = 75 days as shown in this tabulation:

% Biomass Retention	Start Up Time-Days
10	700
20	280
30	160
40	107
50	75
75	40
100	25

Unit processes and techniques which have been used to provide improved solids removal are:

1) Membrane Filtration
2) Granular Media Filtration

Dorr-Oliver offers a membrane solids separation process, but indicates it is only economical on high strength waste waters. Crossflow microfiltration has been successfully used for solids separation in aerobic processes in Japan, but when applied to anaerobic processes, undue plugging is reported. Dynatech has used a pilot scale, pressurized reactor with membrane solids separation for anaerobic conversion of hydrogen, carbon monoxide and carbon dioxide to methane. Alcohol production schemes have employed hollow fiber techniques for cell immobilization.

Dissolved air flotation has reportedly been successfully used for over 20 years for solids separation from an anaerobic contact process treating meat packing plant wastewaters in Minnesota. Apparently the natural gas evolution from the mixed liquor is adequate to provide the gas bubbles for flotation.

Considerable advances have been made in granular media filtration of water. Moving bed filters which are continuously cleaned have been developed to successfully filter water with high suspended solids concentrations. There is the potential for adapting such a process to cell separation in anaerobic treatment. Upflow filtration through granular media increases the capacity for solids removal and prevents plugging beyond a head loss comparable to the submerged weight of the media. It appears promising that adaptation of existing technology for solids separation can be practically adapted to anaerobic treatment, resulting in more efficient solids retention, higher biomass/loading rate characteristics, and improved process stability. I consider this the most crucial developmental need for exploitation of anaerobic treatment.

ENHANCED SPECIFIC ACTIVITY

It is difficult to assay the specific activity of anaerobic processes being fed particulate slurries such as domestic sludge, because the actual biomass cannot be distinguished from the refractory particulates. However, with soluble feedstocks, essentially all of the suspended solids within the reactor is of biological synthesis origin. Commonly the specific activity of prototype anaerobic processes treating soluble industrial wastewaters is approximately 1 kg COD utilized/kg biomass-day. For the biomass converting acetate to methane, there are two classes of methanogens-Methanothrix and Methanosarcina. The first class has a low specific activity and a low half velocity constant which causes it to predominate in systems with low steady state acetate concentrations. This is true of most operating systems because they are lowly loaded. The second class - Methanosarcina - have relatively high half velocity constants and specific activities. In addition, there is a growing body of evidence to indicate their nutritional requirements are more difficult to satisfy.

In highly loaded systems, it has been consistently demonstrated that a shift in predominance from the low rate Methanothrix to the high rate Methanosarcina can be effected by supplementation of trace nutrients. Even though the acetate concentration is high enough to favor the shift in predominance - if only kinetics were involved, no such shift is observed unless nutrients are supplemented. Thus it appears that both nutrient requirements and kinetics control the shift in predominance.

The specific activity of the Methanosarcina has been observed to be three to five times as high as Methanothrix. Acetate to methane conversion rates in 20 day SRT/CSTR reactors were found to be as high as 90 kg/m^3-day when Methanosarcina predominated as compared to about 5 kg/m^3-day when Methanothrix predominated. In addition, Methanothrix systems fail by washout at SRT of 10 days, while Methanosarcina systems can convert 35 kg/m^3-day of acetate to methane at as low as five days SRT.

Dramatic results have also been noted from short term supplementation of trace nutrients. Wilkie, et al (1986) noted rapid declines in what were previously pernicious, high volatile acids concentrations in Napier grass digesters when nickel, cobalt, molybdenum selenium and sulfate were supplemented. Ghosh (1986) noted similar results in the digestion of Bermuda grass. A starch plant anaerobic process in Ireland showed a rapid and dramatic stimulation from the supplementation of trace metals. A municipal sludge digester with elevated volatile acids concentration showed a rapid decline after iron supplementation. A "stuck" digester in Baltimore had a 167% increase in gas production the day after iron was supplemented even though the soluble iron content was very high at 12 mg/l.

Iron, cobalt, nickel, molybdenum, selenium, calcium, magnesium and ppb levels of vitamin B12 have been proven to cause dramatic stimulation in methane production rates. Their relative costs have been minor, but their impact has been great.

SUMMARY

Further developments in anaerobic treatment are postulated to be forthcoming in two major areas. The first area is biomass separation/retention. The second area is careful attention to specific trace nutrient supplementation. It is not overly optimistic to project a quantum increase in the loading rates and stability which will result and such advances could be realized within the next five years.

ACKNOWLEDGEMENT

This research on trace nutrient stimulation of methogenesis was supported by Argonne National Laboratories Contract No. 31-109-38-7196, with C. V. Pearson as the contract officer and also by U.S. Environmental Protection Agency Grant No. R-810633-01-0.

REFERENCES

SUIDAN, MAKRAM, (1986) Personal Communication, University of Illinois-Urbana.

LOWENTHAL, R., DOLD, P. and MARAIS, G. (1986) "An Hypothesis for Pellitization in Upflow UASB Reactor Processes" Proc. of Symposium-Anaerobic Digestion. Bloemfontein, South Africa Sept. 21-24, 1986.

WILKIE, A., GOTO , M., BORDEAUX, F.M. and SMITH, P.H. (1986) "Enhancement of Anaerobic Methanogenesis from Napiergrass by Addition of Micronutrients" BIOMASS In Press.

ANAEROBIC TREATMENT OF PHARMACEUTICAL FERMENTATION WASTEWATER

W.M. Robertson and R.E. Green

Abbott Laboratories
North Chicago, Illinois

ABSTRACT

The anaerobic degradation of pharmaceutical antibiotic fermentation wastewaters was studied at pilot scale, and a system was selected for full scale implementation. The waste contained a high proportion of suspended solids which represented about 40% of the COD, as well as residual amounts of antibiotics, extraction solvents, grain flours, sugars, protein and nutrients. Four treatment configurations were piloted: a downflow anaerobic filter, a downflow/upflow anaerobic filter, an upflow anaerobic sludge blanket, and a low rate anaerobic reactor. The high rate systems were ultimately incapable of assimilating the feed suspended solids, resulting in excessive loss of biomass and, therefore, low soluble COD removals. The low rate system adequately hydrolyzed the feed solids and yielded 70% COD and 80-90% TSS removals. The presence of antibiotic residuals did not affect the system. A full scale low rate reactor system was designed and is planned for startup in August, 1987.

PHARMACEUTICAL INDUSTRY WASTE TREATMENT

Many variations of anaerobic system design have been employed on a wide range of industrial wastewaters. However, there has been comparatively little reported information concerning the amenability of pharmaceutical industry process wastes to direct anaerobic treatment. This is certainly due in part to the extreme diversity of products and processes employed throughout the industry, making it difficult to characterize the resulting wastewaters on an industrywide basis. Anaerobic treatment is also still regarded by many as a technology with a history of inexplicable upsets and occasional process failures and, therefore, many times is considered for use by the design engineer only after the technology's effectiveness has been previously demonstrated on another very similar wastewater.

In its survey of 466 pharmaceutical plants the United States Environmental Protection Agency was able to subcategorize the pharmaceutical industry based on unit manufacturing processes and some common wastewater characteristics (Ref. 1). The subcategories selected were (a) fermentation, (b) biological and natural extraction, (c) chemical synthesis, (d) formulation and (e) research. Wastewaters produced within any one particular subcategory share many common characteristics among different manufacturing facilities and allow one to generalize to a greater extent concerning typical wastewater makeup and treatability. For example, the USEPA survey determined that fermentation wastewaters are typically characterized by high BOD, COD and TSS, large flows and a pH range of about 4.0 to 8.0. Such wastewaters contain small amounts of organic extraction solvents and significant quantities of residual foodstuffs such as grain flours and oils, sugars, starches, protein and macro- and micronutrients and, therefore, are typically treated by biological methods. The USEPA survey also found that, of the 76 biological treatment processes encountered within the pharmaceutical industry, all but two consisted of aerobic processes such as activated sludge, aerated lagoons, trickling filters and stabilization ponds. The remaining two processes were not specified.

ABBOTT WASTEWATER TREATMENT EXPERIENCE

Abbott Laboratories' North Chicago, Illinois facility primarily conducts fermentation and synthesis operations in the manufacture of pharamaceutical preparations. Fermentation processes are employed to produce antibiotics such as erythromycin and various biologically derived pesticides. Wastewater treatment consists of a high rate activated sludge (AS) pretreatment system which discharges to a publicly owned treatment works. Abbott's current average influent and effluent characteristics and pretreatment system performance are presented in Table 1.

Table 1. Activated Sludge (AS) Plant Operating Parameters

Influent	
Flow, m^3/d (mgd)	2930 (.775)
BOD_5, Kg/d (lb/d)	10500 (23200)
TSS, Kg/d (lb/d)	3340 (7340)
Effluent	
Flow, m^3/d (mgd)	2930 (.741)
BOD_5, Kg/d (lb/d)	580 (1280)
TSS, Kg/d (lb/d)	910 (2000)
Waste Activated Sludge	
Flow, m^3/d (mgd)	130 (.034)
BOD_5, Kg/d (lb/d)	890 (1960)
TSS, Kg/d (lb/d)	5500 (12100)
Operating Conditions	
HRT, d	1.4
SRT, d	7.7
F:M	.31
Cell Yield	.34
MLTSS, mg/L	9900
Temperature, C (F)	30 - 40 (85 - 105)

At the Abbott facility the fermentation raw waste stream constitutes 68% of the BOD load to the AS plant, while the chemical sysnthesis raw waste constitutes the remaining 32%. Fermentation production throughout the pharmaceutical industry and at Abbott, as well, is generally continuous and results in the generation of wastewater with relatively consistent and predictable characteristics. Chemical synthesis production, on the other hand, is usually conducted batchwise and in "campaigns", resulting in the generation of a complex wastewater which can vary considerably in constituents and flow rate and which can sometimes be incompatible with biological treatment systems. Because it was known that anaerobic treatment processes function most reliably on waste streams which do not exhibit sudden changes in component makeup, it was decided to evaluate the applicability of anaerobic technology to only the fermentation wastewaters. At Abbott this waste has the characteristics listed in Table 2. The wastewater also contains residual levels of antibiotics which did not affect the performance of the AS pretreatment process, but which were of potential concern in anaerobic treatment considering the slower organism growth rates and the balance among the acid formers, acetogens and methanogens which must be maintained.

Table 2. Fermentation Wastewater Characteristics

Parameter	Value	
Flow, m^3/d (mgd)	1730	(.457)
COD, Kg/d (lb/d)	15000	(33000)
BOD, Kg/d (lb/d)	7150	(15700)
TSS, Kg/d (lb/d)	2880	(6340)
TKN, Kg/d (lb/d)	430	(940)
NH_3-N, Kg/d (lb/d)	280	(620)
P, Total, Kg/d (lb/d)	80	(180)
Sulfate, Kg/d (lb/d)	350	(760)

ANAEROBIC SYSTEM EVALUATION

In 1984 Abbott began the evaluation of anaerobic pretreatment of its process wastewaters ahead of its existing AS process, driven by potential cost savings anticipated through reductions in waste sludge generation and aeration requirements in the AS plant. It is interesting to note that biogas generation was not a significant factor in the analysis, since Abbott's utility boilers are coal fired, resulting in a low BTU cost against which the value of the biogas was measured. A preliminary economic analysis was first prepared based on a hypothetical system design. Although the analysis included many assumptions, the projected payback range of 1.3 to 2.7 years encouraged the initiation of bench and pilot scale studies. After a general review of the literature and consultation with various design/engineering firms, four anaerobic processes were selected for evaluation: (1) a downflow anaerobic filter, (2) a combination downflow/upflow anaerobic filter, (3) an upflow anaerobic sludge blanket (UASB), and (4) a low rate anaerobic reactor. Bench or pilot scale units of each design were constructed and seeded with material obtained from a properly operating anaerobic digester. The units were supplied daily with raw wastewater obtained directly from the fermentation process sewer at the Abbott facility.

Anaerobic Filters

The anaerobic filters were selected first for pilot testing, since attached growth systems offer a potentially long solids retention time (SRT) which promotes the establishment of a highly acclimated biomass. These systems are generally regarded as tolerant of load swings or brief toxic slugs, since only the surface layer of the biomass attached to the support media is directly exposed to the wastewater. The downflow filter consisted of commercial biofilter media enclosed in a clear, plastic column and fitted with appropriate liquid and gas connections. The downflow/upflow filter was constructed by placing two identical columns in series so that liquid flow was downward in the first column and upward in the second. The units were seeded with a healthy municipal digester sludge and gradually loaded with fermentation wastewater. Performance of both of the anaerobic filters initially appeared to be good (Tables 3 and 4). As the loadings were increased, however, both units experienced upsets, followed occasionally by complete reactor failure. The most common upset exhibited a sharp decrease in pH concurrent with a rise in volatile acids and a loss of gas production. At times the biomass also experienced detachment from the support media, requiring shutdown and reseeding. In addition the feed suspended solids accumulated within the filters, causing plugging of sections of the support media and resulting in channeling and a lower effective hydraulic retention time (HRT). The downflow/upflow filter was more tolerant of feed solids and achieved higher loadings before performance deteriorated. Based on further analysis of the wastewater it was noted that approximately 40% and occasionally up to 50% of the total COD in this wastewater was associated with the suspended solids. Because the short hydraulic retention time of the anaerobic filters did not allow sufficient solubilization of the solid matter to occur, it was concluded that the observed COD removal was mostly caused by entrapment of suspended solids and was not primarily attributable to the conversion of COD to methane as desired. This conclusion was further supported by the unstable and very low rate of gas generation within either filter unit.

Table 3. Downflow Anaerobic Filter Summary

Media depth, m (ft)	1.2 (4.0)
HRT, d	2.5
Temperature, C (F)	35 (95)
CH_4 Yield, m^3/Kg COD	Unstable

Loading Kg COD/m^3-d	Total COD Removal %	TSS Removal %
1.15	0	87
1.38	55	79
1.54	48	90
1.60	69	96 (System failed)

Table 4. Downflow/Upflow Anaerobic Filter Summary

Media depth, m (ft)	2.4 (8.0)
HRT, d	2.5
Temperature, C (F)	35 (95)
CH_4 Yield, m^3/Kg COD	Unstable

Loading, Kg COD/m^3-d	Total COD Removal %	TSS Removal %
1.0	58	86
2.1	47	94
3.9	62	85
5.3	2	90
5.5	10	68 (System failed)

Upflow Anaerobic Sludge Blanket Reactor

Observation of the downflow/upflow filter revealed that an active sludge blanket had formed in the upflow column and that biomass formation on the upflow media was incomplete, suggesting that the improved performance of the unit compared to the downflow filter may have been attributable to the presence of the sludge blanket, rather than to the presence of the additional upflow media. Because it was known that the upflow anaerobic sludge blanket (UASB) process had been successfully employed full scale on yeast production, beer fermentation and food processing wastes, which were felt to be similar to pharmaceutical fermentation wastes, the application of the UASB technology to this waste was discussed with a UASB process supplier. The supplier indicated that the pharmaceutical waste contained a higher level of suspended solids than was generally recommended for treatment by the UASB process but agreed to assist with pilot testing, because the previous experience with the downflow/upflow filter indicated that the raw waste suspended solids may be trapped in the sludge blanket for a sufficient period to allow solubilization to occur.

A bench scale UASB system was obtained and operated for approximately nine months. As indicated in Table 5, the UASB performance was substantially better than either of the anaerobic filter systems, and loadings up to 8 Kg COD/m^3-d were demonstrated to yield satisfactory COD removals and gas production. The target design loading for this system was 8 to 10 Kg COD/m^3-d. However, after extended operation at near design loadings the UASB configuration was also ultimately unable to tolerate the high levels of suspended solids present in the feed. As the loading rate was increased, feed solids collected in the sludge blanket and solids separator sections at a greater rate than the system could accomodate, resulting in excessive bed expansion and the loss of biomass and solids through the column effluent. In addition it became necessary to increase the sludge wasting rate in order to maintain the proper sludge blanket depth, thus further depleting the amount of active biomass present in the system. At loadings less than about 5 Kg COD/m^3-d, however, the UASB process demonstrated consistent COD removals of 70% - 75% on this wastewater without indication of excessive solids accumulation.

Table 5. Upflow Anaerobic Sludge Blanket Summary

Reactor depth, m (ft)	1.2 (4.0)
HRT, d	.48 - .52
Temperature, C (F)	35 - 37 (95 - 98)
CH_4 Yield, m^3/Kg COD	.18 - .30

Loading, Kg COD/m^3-d	COD Removal %	TSS Removal %
1	24	30
3	61	51
5	73	75
8	74	85
10	68	0 (System failed)

Low Rate Anaerobic Reactor

Because the raw waste suspended solids constituted a major portion of the COD in this wastewater, and because the high rate systems which were evaluated did not satisfactorily degrade these solids in the reactor time available, a low rate anaerobic system was considered. Low rate systems, as the term implies, operate at a slightly lower temperature, degrade the waste at a lower rate and are loaded much less heavily than the more common high rate systems, such as those described above. As a result the reactor volume is typically large and offers a solids retention time (SRT) of one to two years, which is considerably greater than that achieved in high rate systems and which is sufficient to permit nearly complete hydrolysis and degradation of raw waste suspended solids. A low rate system supplier was thus retained to provide pilot testing equipment and supervision.

The pilot system was seeded with municipal digester sludge and supplied daily with fermentation wastewater. The target design loading for this system was 1.0 Kg COD/m^3-d. Because the HRT of the low rate system was about 10 days, the COD loading was increased during startup only after at least 10 days had passed since the previous increase and only if effluent volatile acids, gas production, alkalinity and pH remained steady. The COD load applied to the reactor was increased initially from .05 to .40 Kg COD/m^3-d during about the first 50 days of system operation and held at .40 Kg COD/m^3-d for about 30 days. The COD load was then further increased in gradual steps over 40 days to the design load of 1.0 Kg COD/m^3-d. At loadings of 0.84 Kg COD/m^3-d and above, return activated sludge from the onsite AS plant was added to the reactor feed and included in the loading calculations, since the full scale anaerobic system was intended to act as a pretreatment stage ahead of the AS plant, and since all sludge wasting was desired to occur from the anaerobic reactor. COD removal efficiency was maintained by controlled reactor mixing to facilitate contact between the waste feed and the anaerobic biota.

The low rate pilot unit demonstrated stable operation from startup through full loading. Although the presence of residual erythromycin antibiotic material in the feed wastewater was monitored daily by bioassay techniques, the material was not injurious to the system's performance. Furthermore, analysis of the anaerobically treated effluent revealed that the antibiotic was substantially degraded within the system. The information presented in Table 6 reflects over seven months of pilot plant operation at this time.

Table 6. Low Rate Anaerobic Reactor Summary

Reactor length, m (ft)	3.3 (10.7)
Reactor diameter, m (ft)	0.6 (2.0)
HRT, d	9.5 - 10.5
Temperature, deg C (F)	28.5 - 32.5 (83 - 90)
CH_4 Yield, m^3/Kg COD	.25 - .31

Loading Kg COD/m^3-d	Total COD Removal %	TSS Removal %
.09	62	85
.20	71	85
.38	72	80
.65	59	88
.84 (incl RAS)	69	94
1.03 " "	70	93
1.10 " "	70	83

FULL SCALE IMPLEMENTATION

Because of its stable operation and ability to handle raw waste solids as well as return activated sludge, the low rate anaerobic treatment process has been selected for full scale implementation at Abbott's North Chicago facility. The full scale design performance is presented in Table 7. Figure 1 illustrates the treatment process configuration as it will ultimately be employed. As the diagram indicates, the fermentation raw waste stream will be separately pretreated through two low rate reactors operating in parallel before joining the chemical synthesis raw waste stream. The combined wastewater will then flow to the existing activated sludge plant for final treatment. Waste activated sludge will be returned and mixed with the anaerobic reactor feed to provide complete solids digestion, and final sludge wasting will take place from the reactors. Biogas will be directed to an existing boiler on site which has both coal and gas firing capability. Startup is expected in August, 1987, with full loading to be reached about November, 1987.

Table 7. Full Scale Design Removal Performance

	Low Rate Reactor Only	Complete System
COD, Total	65%	90%
BOD_5, Total	65%	95%
TSS	80%	80%

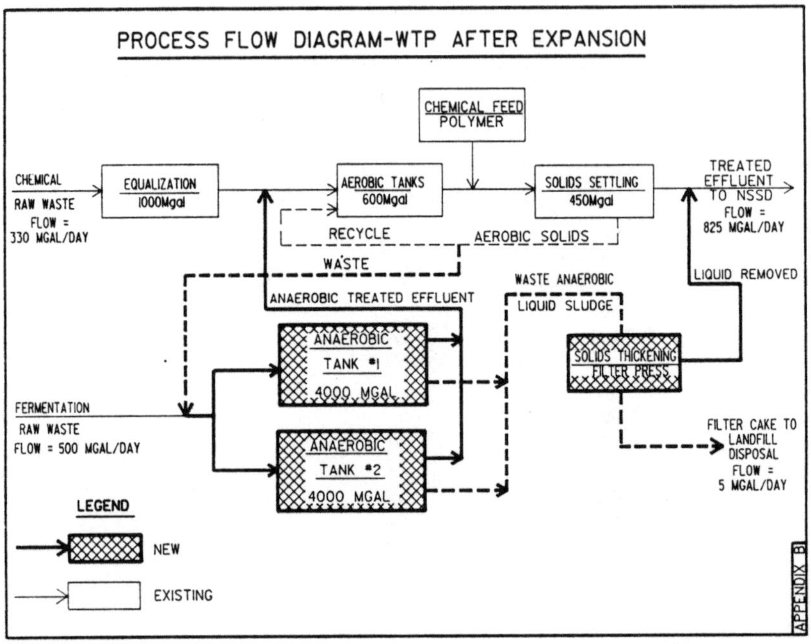

Figure 1. Full Scale Process Flow Schematic

REFERENCES

1. U. S. Environmental Protection Agency, Development Document for Final Effluent Limitations Guidelines and Standards for the Pharmaceutical Manufacturing Point Source Category, EPA 440/1-83/084, September, 1983.

ANAEROBIC TREATMENT FOR PULP AND PAPER WASTEWATERS

E.R. Hall, H. Melcer, L.-A. Cornacchio, and R.M. Jones

Environment Canada
Wastewater Technology Centre
Burlington, Ontario, Canada

ABSTRACT

Anaerobic technology has been gaining acceptance as a cost-effective treatment alternative for wastewaters produced in the pulp and paper industry. Potential applications for anaerobic treatment are being delineated by an extensive laboratory scale screening study of several Canadian pulp and paper wastewaters. These testing procedures estimate the level of COD and BOD removal achievable anaerobically, as well as the effects of inhibitory wastewater components on anaerobic microorganisms. Results of a pilot scale comparative assessment of several commercial anaerobic systems are reviewed along with data illustrating the effects of dynamic operating conditions on the stability of high rate anaerobic processes. The preliminary full scale process design derived from this study is also presented, with estimates of the associated capital and operating costs. In addition, a summary is provided of the other major applied anaerobic development programs in Canada. Finally, the priority ratings for applications of anaerobic technology in the pulp and paper sector are discussed along with an indication of the major unresolved technical issues associated with these applications.

INTRODUCTION

Between 1969 and 1980, expenditures by the Canadian pulp and paper industry for pollution abatement amounted to about C$ 660 million. As a result of these measures, in 1980, 72% of Canadian mills were in compliance with federal regulations governing the discharge of BOD (biochemical oxygen demand) in liquid effluents (1). One approach to further improve the compliance status involves changes in manufacturing technology to processes such as thermo-mechanical pulping (TMP). By increasing product yield with mechanical pulping, the total amount of pollution discharged is reduced. However, by implementing these changes in conjunction with decreased water usage, the final effluents discharged are often much more concentrated.

Anaerobic processes are widely used for the treatment of food and beverage industry wastewaters. The trend to higher concentration effluents in a number of pulp and paper mills has led to an increased awareness of the potential for treating these wastewaters in a cost-effective mannner with anaerobic technology (2).

Anaerobic methane fermentation technology offers many potential advantages for the treatment of concentrated industrial organic wastewaters. In the case of the pulp and paper sector, the usual benefits of reduced treatment costs and energy recovery may be expected together with improvements in treatment efficiency and toxicity reduction. Basic research completed in

Scandinavia suggests that anaerobic processes may provide enhanced degradation of chlorinated organics from bleach plants as well as improved removal of reduced sulfur compounds from spent pulping liquors and evaporator condensates.

Before adopting anaerobic treatment for these applications, the industry requires additional information on several aspects of the technology (Table 1). This paper illustrates the activities that are currently underway to generate these data, by reviewing Canadian programs involving anaerobic treatment of pulp and paper wastewaters.

Table 1. Information Required for the Implementation of Anaerobic Treatment in the Pulp and Paper Industry

1. General Suitability of process effluents.
    - strength
    - flow
    - temperature
    - suspended solids

2. Biodegradability of Wastewater Organics.
    - treatment efficiency
    - methane production

3. Presence and Effects of Toxic or Inhibitory Compounds.

4. Process Selection, Design and Economics.

SUITABILITY, BIODEGRADABILTY AND TOXICITY OF PULP AND PAPER WASTEWATERS

In 1985, Environment Canada's Wastewater Technology Centre (WTC) initiated a program to provide assistance in the implementation of anaerobic treatment technology in the Canadian pulp and paper industry (3). To assess the scope for applications of anaerobic treatment in Canada, 21 pulp and paper mills and 42 in-plant waste streams were selected for screening at the WTC (Table 2). These mills were representative of the variety of raw materials and manufacturing technologies currently used in the Canadian industry. The screening program involved chemical characterization using conventional methods as well as biological testing using an anaerobic serum bottle technique (4). The serum bottle procedure allowed a relatively rapid assessment of several anaerobic treatability characteristics such as COD and BOD removal, biogas production, and the inhibitory properties of the wastewater on anaerobic microorganisms.

Data typically produced during these tests are shown in Figure 1 which illustrates the contrasting results obtained from serum bottle testing of two pulp and paper wastewaters. The cumulative methane production patterns on the left show no initial lag periods, and consistent increases in total methane produced, with increasing wastewater concentrations tested. These

Table 2. Summary of Mills and In-plant Streams Selected for Anaerobic Treatability Screening

P&P Technology	In-Plant Streams
1. KRAFT PULPING	
Unbleached Kraft	Combined Mill Effluent
Bleached Kraft	Woodroom Effluents
	Weak Black Liquor
	Bleach Plant Caustic Extracts
	Bleach Plant Chlorination Filtrate
	Evaporator and Digester Condensates
2. SULFITE PULPING	
Sulfite & Groundwood	Spent Liquor
Ammonium Base Sulfite	Combined Mill Effluent
High Yield Sulfite	Spent Liquor
Very High Yield Sulfite	Spent Liquor
Neutral Semi-chemical	Spent Liquor
3. MECHANICAL PULPING	
CTMP and NSSC	Combined Mill Effluent
Peroxide Bleached TCMP	Combined Mill Effluent
CMP	Spent Liquor
TMP and CTMP	Combined Mill Effluent
TMP	Chip Wash Effluent
	Combined Mill Effluent
4. MISCELLANEOUS PULPING	
Non-Sulfur Pulping	Combined Mill Effluent
	White Water Recycle
Masonite Hardboard	Combined Mill Effluent

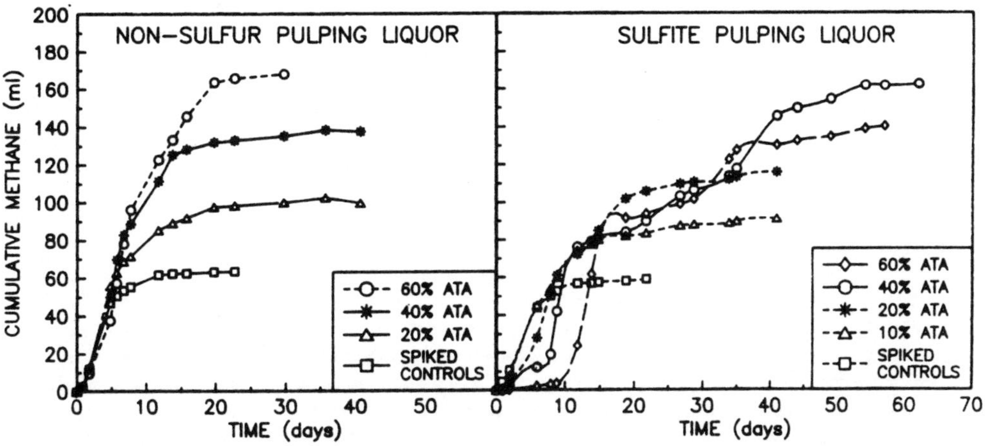

Figure 1. Selected Anaerobic Toxicity Assay Results

results indicate that the non-sulfur pulping liquor should be readily treatable anaerobically without inhibition of the anaerobic biomass. On the other hand, the sulfur-based pulping liquor demonstrated an inhibitory effect which may be attributed to the presence of significant quantities of organic and inorganic sulfur. As shown on the right of Figure 1, at higher concentrations of wastewater, the lag periods in the onset of methane production were longer. However, once methane production began, the rate of accumulation was similar to that of the control, indicating that acclimation to the inhibitory components is possible. The relatively complex nature of the methane production curves for this wastewater also suggest sequential removal of several biodegradable constituents.

It is anticipated that the results of the screening program will assist in the establishment of priorities for a more detailed R&D program. This would include identification of wastewaters which are suitable for immediate treatability or demonstration studies, wastewaters which warrant additional research to improve their anaerobic treatabilities, and wastewaters which are unsuitable for anaerobic treatment.

PROCESS SELECTION AND DESIGN

From laboratory scale batch data such as those illustrated by Figure 1, several pulp and paper wastewaters were found to be suitable for immediate pilot scale treatability assessment. One example was a relatively small mill owned by MacMillan Bloedel at Sturgeon Falls, Ontario. At this site, 200 tonnes/day of corrugating medium is produced in a neutral sulfite semi-chemical (NSSC) process, and 100 tonnes/day of hardboard siding is manufactured from mechanical pulp. The combined effluents are discharged to the Sturgeon River, after primary treatment for suspended solids removal.

Since little information was available on the anaerobic treatment of spent pulping liquors, a cooperative research program involving MacMillan Bloedel, Environment Canada, and the National Research Council of Canada was initiated in 1984 to investigate the application of anaerobic technology to the wastewater at Sturgeon Falls. Under Task I of the program, a preliminary assessment of the six major plant sewers was carried out at the University of Washington, using serum bottle testing protocols. These results indicated that all in-plant waste streams were amenable to anaerobic treatment. However, some inhibition was associated with the NSSC spent liquor stream. Therefore, it was recommended that the combined mill effluent be considered as the most promising candidate for further study since the spent liquor was diluted by 60% in this stream by the other mill wastewaters.

Task II was initiated at the WTC in June 1984. This phase involved the evaluation of five different pilot scale anaerobic reactors with Sturgeon Falls effluent. Three of the processes evaluated - an upflow anaerobic sludge blanket (UASB) reactor, a fluidized bed, and a hybrid sludge blanket/anaerobic filter - are shown schematically in Figure 2. Several of the pilot reactors were operated successfully at organic loadings as high as 25 kg COD/m^3·d, corresponding to hydraulic retention times of less than 1 day. Table 3 indicates selected results from the experimental program. Treatment efficiencies of 50-60% and 80-90% for COD and BOD respectively were achieved at loadings of 15-16 kg COD/m^3·d with the addition of small quantities of nitrogen and phosphorus (5). When the experimental program was completed in January 1986, several successful technical alternatives were identified for demonstration at the mill.

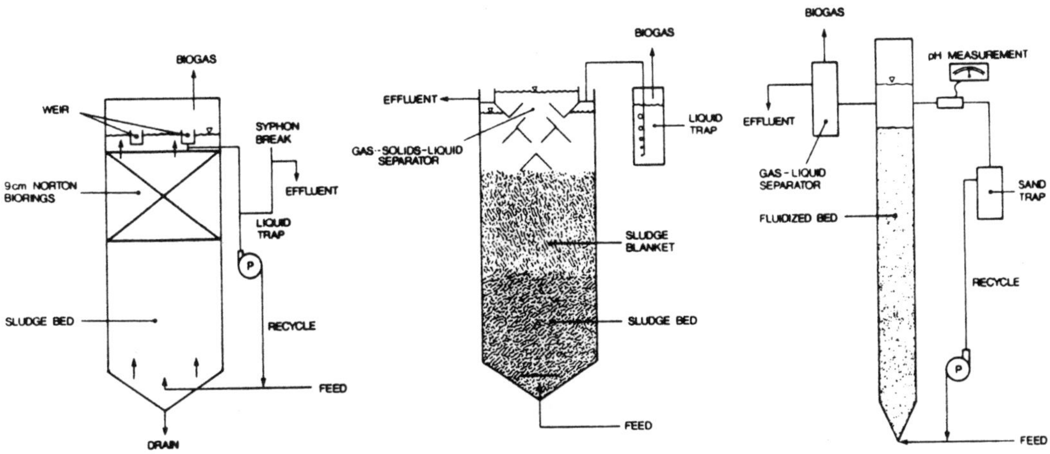

Figure 2. Hybrid, Sludge Blanket (Biopaq) and Fluidized Bed (Anitron) Reactors

Table 3. Performance Summary for Hybrid, Biopaq and Anitron Reactors (5).

Reactor	Influent COD mg/L	Organic Loading kg COD/m^3·d	% Removal BOD	% Removal COD
HYBRID	22 100	16.2	75	54
BIOPAQ	12 700	14.5	85	56
BIOPAQ	19 500	15.3	91	58
ANITRON	12 700	13.2	81	54
ANITRON	18 940	16.2	90	58

After discussions with several suppliers of anaerobic treatment systems, MacMillan Bloedel selected the UASB process offered by Paques Lavalin Ltd. of Toronto, on the basis of its technical performance and cost. The full scale plant proposed by Paques Lavalin was designed and costed (Table 4) assuming an organic loading rate of 10 kg COD/m^3·d and a retention time of 2 days. The demonstration plant being utilized in Task III consists of a 20 m^3 sludge blanket reactor with a pre-acidification stage and facilities for nutrient addition and pH control. Construction and installation of the demonstration plant was completed in April 1986 and on May 9, the reactor was seeded with granular sludge from a full scale UASB in the Netherlands and wastewater feeding was initiated.

Table 4. Design and Costs of Full Scale Biopaq Upflow Anaerobic Sludge Blanket Reactor (5).

TECHNICAL		
Reactor Volume	:	13 000 m^3 (6 reactors)
COD Loading	:	130 000 kg/day
BOD Loading	:	50 000 kg/day
Treatment Efficiency	:	50% COD removal >75% BOD removal
Biogas Production	:	20 000 m^3/day
ECONOMIC		
Capital Cost	:	C$ 6 000 000
Annual Biogas Value	:	C$ 900 000
Annual Operating Cost	:	C$ 500 000

The design loading rate of 10 kg COD/$m^3 \cdot d$ was reached after 1 month of operation. A one-month steady state run was then completed to define the system performance at the design load. Although the average loading was maintained at approximately 10 kg/$m^3 \cdot d$, the feed rate was allowed to vary in proportion to the total mill effluent flow. In subsequent runs, the UASB reactor has been operated in a similar manner, but with loading rates of up to 20 kg COD/$m^3 \cdot d$ to determine the feasibility of decreasing the size of the full scale facility. Under all loading rates examined to date, the target BOD removal rates of 80% have been exceeded.

Several other pilot projects are currently in progress in Canada. Table 5 lists these programs and indicates the unique aspects of each study. In every case, the use of anaerobic technology at these locations requires that additional research be completed to address specific technical issues that are new to anaerobic industrial wastewater treatment. At Sturgeon Falls and Bathurst, two novel high rate anaerobic reactor designs are also being studied (internal circulation reactor and Gist-brocades fluidized bed).

Table 5. Canadian Pilot and Demonstration Scale Studies, October 1986.

1. STURGEON FALLS, ONTARIO (NSSC & Hardboard)
   - UASB and Internal Circulation (IC) Reactor
2. BATHURST, NEW BRUNSWICK (CTMP & NSSC)
   - UASB, Anaerobic contact and Fluidized bed
   - Peroxide bleaching
3. CABANO, QUEBEC (NON-SULFUR PULPING)
   - UASB
   - Discharge to Municipal WPCP
4. KAPUSKASING, ONTARIO (TMP & Magnefite)
   - Hybrid reactor
   - Thermophilic Anaerobic Treatment
5. QUESNEL RIVER, B.C. (CTMP & TMP)
   - Alternating CTMP and TMP Wastewater

These are projected to operate at loading rates that are 100% to 300% higher than those currently used in full scale anaerobic treatment facilities.

The projects shown in Table 5 are indicative of the current priority applications in the pulp and paper industry. In Canada, anaerobic treatment may be attractive for older, smaller sulfite mills that do not have chemical recovery systems, or, for newer mills that employ variations on high yield mechanical pulping. Both of these mill types produce relatively high strength combined effluents. For mills with recovery systems, anaerobic treatment could be used for selected wastewater streams. For example, bleach plant effluents and weak black liquor could be treated anaerobically to off-load components of the wastewater treatment and chemical recovery systems. In a similar fashion, anaerobic treatment could be integrated into an existing wastewater treatment plant to upgrade the performance of an aerobic process, or to accommodate an expansion of production.

RESEARCH ISSUES IN THE PULP AND PAPER INDUSTRY

Although a number of anaerobic pilot scale studies have already been initiated in the Canadian pulp and paper sector, there are also several unresolved technical issues that are being addressed by ongoing and planned research programs. As with many non-food industry wastewaters, pulp and paper effluents contain variable amounts of materials that are toxic or inhibitory to microorganisms in a biological treatment plant. The effects of these time-varying factors can be mediated partially by altering process designs to include adequate equalization, recycle and bypass capability. Operating strategies can also be optimized with on-line monitoring and control systems that can respond to process stress by manipulating a number of system variables. Studies at the WTC (6) have shown that high rate anaerobic reactors treating pulp and paper wastewater can resist process upset during transients in operating conditions (Figure 3). Similar

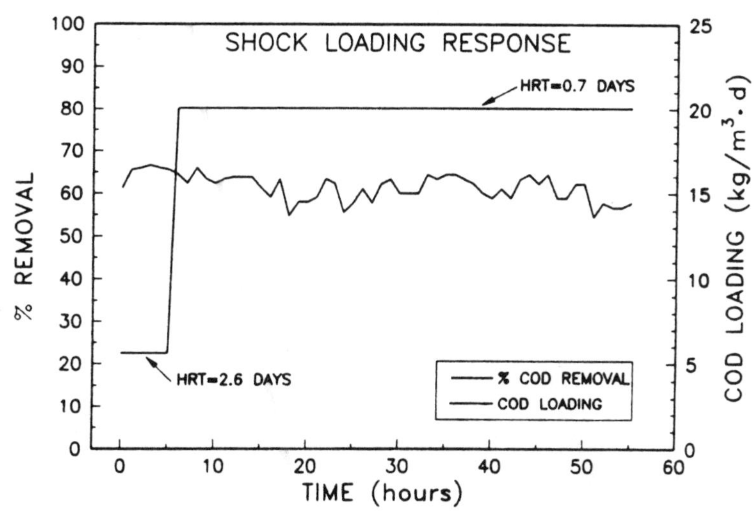

Figure. 3 Response of Pilot Scale Hybrid Reactor to Feed Rate Step

procedures are currently being employed for the development of expert systems that will maintain process integrity under the dynamic operating conditions found at pulp and paper mills.

Canadian pulp and paper mills must generally reduce the aquatic toxicity of effluents that are discharged to receiving waters. The current standard technology for the production of a non-toxic effluent is the 5-day aerobic stabilization basin. The efficacy of anaerobic treatment for the reduction of toxic components of pulp and paper wastewaters is being assessed in another WTC program. The anaerobic-aerobic process conditions required to remove resin and fatty acids, chlorinated organics, and total reduced sulfur compounds, are being examined. The effects of conventional and innovative pulp bleaching processes on anaerobic systems are also of interest. In order to deal effectively with wastewaters from chlorine and peroxide bleacheries, traditional anaerobic treatment plant flowsheets will need to be changed to provide detoxification prior to methane production (7).

REFERENCES

1. Environment Canada, "Status Report on Abatement of Water Pollution from the Canadian Pulp and Paper Industry (1980)", Government of Canada, Department of the Environment Report No. EPS 3-WP-82-3, Ottawa, 1982.

2. Rantala, P. and A. Luonsi (eds.), Proc. of IAWPRC Seminar on Anaerobic Treatment of Forest Industry Wastewaters, Wat. Sci. Tech., 17(1), (1985).

3. Hall, E.R., "Development of Anaerobic Treatment Technology for Pulp and Paper Wastewaters", Environment Canada, Wastewater Technology Centre, Burlington, Ontario, November 1984.

4. Cornacchio, L.-A., E.R. Hall and J.T. Trevors, "Modified Serum Bottle Testing Procedures For Industrial Wastewaters", Presented at 21st Can. Symp. Wat. Poll. Res., Burlington, Ontario, April 30, 1986.

5. Hall, E.R., C.F. Prong, P.D. Robson and A.J. Chmelauskas, "Evaluation Of Anaerobic Treatment For NSSC Wastewater", Presented at 1986 Environmental Conf., TAPPI, New Orleans, April, 1986.

6. Jones, R.M. and E.R. Hall, "Dynamic Modelling For Process Control Of A High Rate Anaerobic Reactor", Proc. EWPCA Conf. on Anaerobic Wastewater Treatment", Amsterdam, September, p. 621, 1986.

7. Welander, T. and P.-E. Andersson, "Anaerobic Treatment Of Wastewater From The Production Of Chemi-thermomechanical Pulp", Wat. Sci. Tech., 17(1), 103, (1985).

ACKNOWLEDGEMENTS

Some of the studies described in this paper were partially supported by the Federal Panel on Energy R&D (PERD).

ANAEROBIC WASTEWATER TREATMENT OF A FUEL ETHANOL FACILITY

Robert L. Gross
Ashland Petroleum Company
Ashland, Kentucky

Jelte Lanting
Biothane Corporation
Camden, New Jersey

ABSTRACT

In an effort to consistently meet an existing NPDES permit a corn to fuel grade ethanol producer in southeastern Ohio recently installed a high rate anaerobic wastewater treatment system. This plant uses the upflow sludge blanket technology to pretreat process wastewater. The effluent from the anaerobic system is polished in a two-stage bio-tower prior to discharge to the Ohio River.

Included in the paper is a brief overview of the corn-to-ethanol process and a description of the types of wastewater generated. The results of a five month treatability study are presented and compared with the operation of the full scale system. Finally, the design criteria, the start-up problems, and the recent operation of the anaerobic system are discussed.

INTRODUCTION

In 1984 South Point Ethanol (SPE), a major ethanol producer in Southern Ohio, reached an agreement with the Ohio EPA to improve the performance of its wastewater treatment facility, the heart of which was a two-stage trickling filter. After evaluating alternatives, South Point Ethanol decided that anaerobic pretreatment of its wastewater would probably be the most cost effective way to solve its compliance problems. To determine the anaerobic treatability of the ethanol facility's wastewater, a contractor was selected to carry out an on-site pilot study using a 6 cubic meter (1600 gls) Upflow Anaerobic Sludge Blanket (UASB) reactor. During the Summer and Fall of 1984, the 21-week pilot study demonstrated that the highly variable wastewater generated at the ethanol facility could be effectively pretreated using an anaerobic system. Furthermore, the results of the attempt to polish the anaerobic effluent in a two-stage pilot trickling filter suggested that a high quality final effluent could be expected from a combined system. Results of this study have been discussed in detail by Lanting and Gross (1985).

Based on the favorable results of the pilot study, SPE proceeded to install a full-scale UASB pretreatment system. This system began operation early in May of 1986 and achieved excellent removal efficiencies under full load conditions in July. This paper will focus on the start-up and steady state performance of the anaerobic system, after a brief review of the pilot study and its implications.

SOUTH POINT ETHANOL

SPE is a corn to fuel grade ethanol facility producing 230,000 cubic meters (approximately 60,000,000 gls) of denatured ethanol and 200,000 metric tons of distillers dried grain with solubles (DDGS) each year. Approximately 100 cubic meters (2800 bushels) per hour corn is pulverized, cooked, mixed with enzymes and yeast, and added to one of eight large fermenters. The fermented corn slurry (beer) is pumped through a series of columns to distill off the product ethanol. Residual corn is dried using centrifuges, dryers and evaporators prior to being sold at DDGS. The facility operates 7 days a week, 330 days a year, with two scheduled maintenance outages during the year.

THE WASTEWATER

The majority of wastewater to be treated, about 90%, is generated in the evaporation and drying processes. The remaining 10% comes from the distillation area and miscellaneous washwater. The wastewater is a highly soluble mixture of weak organic acids such as acetic, propionic, lactic and butyric acid. The organic strength of the wastewater depends on the quality of the fermenters but typically ranges from 3000 to 5000 mg/l SCOD. The wastewater flow averages about 4900 cubic meters per day (1.3 mgd) with maximum rates not exceeding 5500 m^3/d (1.45 mgd).

THE PILOT STUDY

The pilot study was carried out with a 6 cubic meter (1600 gls) skid mounted reactor. During the 21 week study, raw wastewater was continuously added to the anaerobic reactor at rates between 4 and 20 m^3/day. The performance and chemical requirements of the anaerobic system averaged over the entire pilot study have been shown in Table 1. In our opinion these data represent a good design basis for the application of the particular UASB technology to similar wastewaters.

During the final phase of the pilot study two pilot trickling filters were operated in series to study the treatment efficiency of the overall system as it was envisioned to operate under full-scale conditions. Dilute sludge from the existing final clarifier was used to seed the pilot filters. Anaerobic effluent was then pumped through the filters for several weeks to acclimate the organisms. During the final weeks of the pilot study a BOD sampling program was carried out. The results indicated that the combination of an anaerobic system and trickling filters worked extremely well on this type of wastewater. During a one month period of time the anaerobic system effluent had an average soluble BOD of 202 mg/l while the first and second stage filter effluent averaged 41 mg/l and 19 mg/l, respectively.

TABLE 1 UASB Pilot Plant Performance Characteristics

	Influent	Effluent	Removal
TCOD (mg/l)	3627	874	76%
SCOD (mg/l)	2889	416	86%
TBOD (mg/l)	2441	288	88%
SBOD (mg/l)	1910	181	90%
TSS	414	330	
Volumetric Loading	9.3 kg $TCOD/m^3.d$		
Hydraulic Retention Time	9.4 hr		
Methane Content Biogas	83%		
Methane Yield	0.33 m^3/kg TCOD rem.		
	Projected Chemical Requirement		
Caustic	0.080 kg NaOH/kg TCOD influent		
Ammonia	0.005 kg N/kg TCOD influent		
Phosphoric Acid	0.001 kg P/kg TCOD influent		

WASTEWATER TREATMENT PLANT DESIGN

A schematic of the wastewater treatment plant as it existed prior to construction has been shown in Fig. 1. The wastewater from various sources was collected in a 7600 m^3 (2,000,000 gls) surge tank. Wastewater was pumped from the surge tank through heat exchangers to cool the water to an acceptable temperature and then entered the suction side of the recycle pumps of the first stage trickling filter. After biological treatment in the first stage and intermediate clarification, transfer pumps carried the water to the second stage trickling filter. Second stage effluent entered the final clarifier and was then discharged into the Ohio River. Sludge from the intermediate and final clarifiers was transferred to a sludge decant vessel prior to dewatering on a belt press.

From the pilot study it was known that additional equalization, more reliable temperature control and improved pH control would be required for trouble free operation of the full-scale facility. Thus, in addition to the UASB reactor, several other unit processes were added to the WTP area. Since an old gasholder vessel was available at the treatment site, it was decided to put that 7600 m^3 (2,000,000 gls) tank into use as an equalization tank. A primary clarifier with a volume of 410 m^3 (110,000 gls) was installed between the surge tank and equalization tank to reduce the incoming suspended solids load on the system. This also helped to reduce the levels of oil and grease associated with these same solids. A 205 m^3 (54,000 gls) surplus sludge tank was installed to store an on-site supply of acclimated sludge. This would provide sludge to re-seed the anaerobic reactor in the event that a chemical spill or some other unexpected situation resulted in a severe upset to the digester's active organisms.

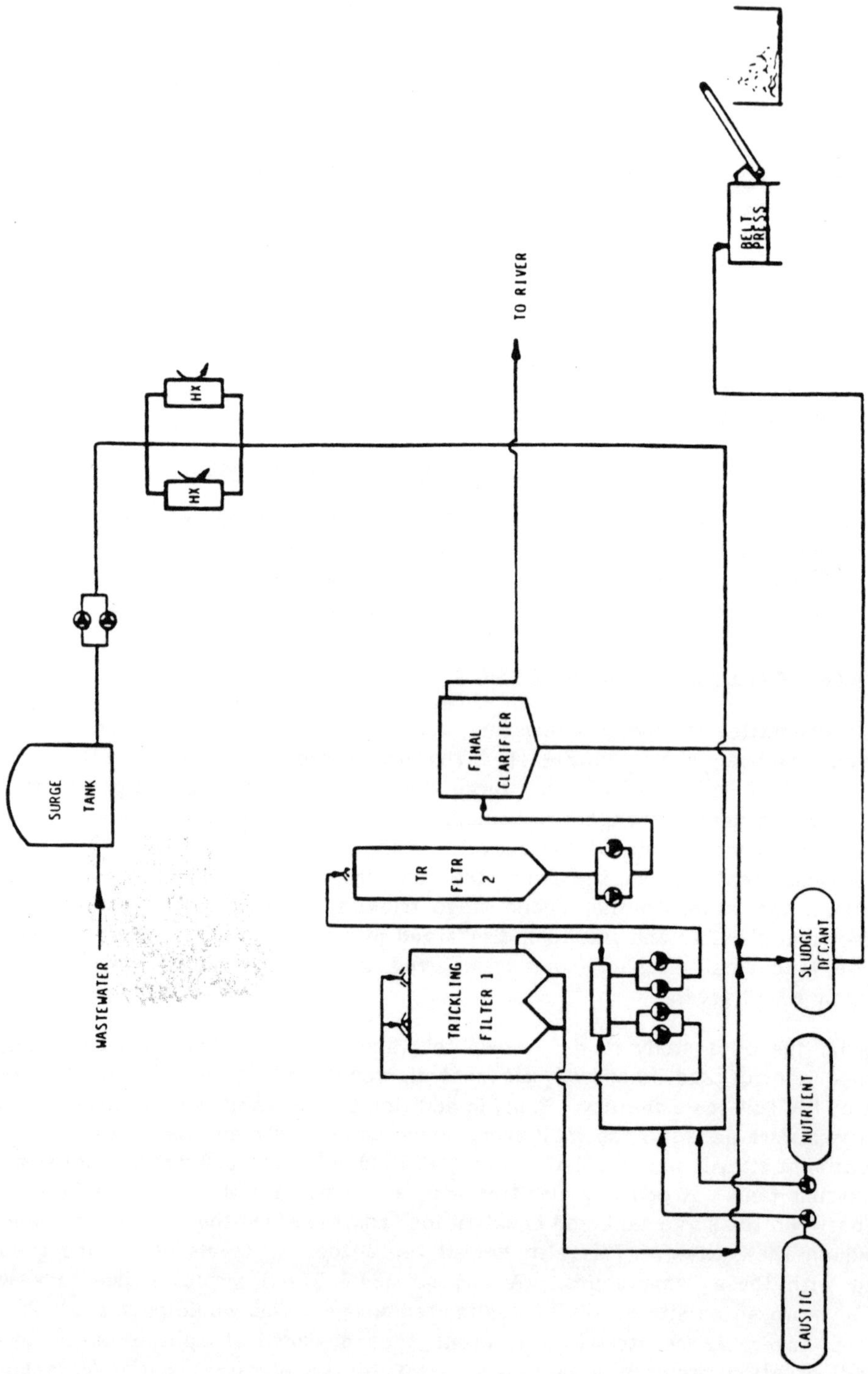

FIGURE 1 Schematic of Treatment Plant Before Construction

Finally, South Point Ethanol installed a 410 m^3 (110,000 gls) sludge thickener, in which sludges from the primary, intermediate and final clarification steps and excess anaerobic sludge were to be mixed and concentrated before dewatering. After construction, which started in November of 1985 and was completed in May 1986, a schematic of the plant now looks as follows (see Fig. 2).

At the heart of the WTP modifications an UASB system consisting of a 47 m^3 (12,500 gls) conditioning tank and 2050 m^3 (540,000 gls) UASB reactor, was inserted between the equalization tank and first stage trickling filter.

After conditioning, the wastewater is introduced into the digester through a network of diffuser pipes located on the floor that ensure even distribution over the entire cross-sectional area of the vessel. Wastewater flows upward through a dense blanket (5-15% solids) of active organisms where the majority of treatment occurs. Treated water, bio-solids and methane gas are separated at the top of the reactor in a patented separator. The separators allow for effective degassing of the biomass and encourage the development of a highly settleable granular biomass. This type of design allows the reactor to handle volumetric loadings in excess of 10 Kg COD per cubic meter of reactor volume per day at COD removal efficiencies ranging from 70 to 95 percent for most applications.

START-UP AND OPERATION

The full-scale facility was started early in May of 1986. A portion of the total wastewater flow (approximately 400 m^3/d) was diverted to the new equalization tank and on through the anaerobic system. The effluent from the anaerobic system was then blended with the remainder of the raw water entering the trickling filter. It was anticipated that after the anerobic system had reached satisfactory levels of stable performance for three or four days that the portion of wastewater diverted to the anaerobic system would be increased another 400 m^3/d. With this type of stepwise procedure, full flow through the anaerobic system was expected within approximately eight weeks. Data describing the plant performance during and following start-up have been graphically presented in Fig. 3.

For start-up the digester had been seeded with 230 m^3 (60,000 gls) of granular biomass from one of the full-scale UASB plants in Wisconsin. This was equivalent to about 10,000 Kg of VSS (22,000 lbs). At the very beginning of start-up the digester was accidentally hit with a slug of high strength wastewater. This occurred during a semi-annual factory turnaround, during which the production facility is shut down for inspection, cleaning and maintenance. From the pilot study it was known that the wastewater generated during a turnaround would stress the biomass. However, the components responsible for the stress symptoms, such as suppressed activity and accumulation of intermediate biodegradation products, have never been identified.

After the turnaround, plant performance slowly improved, but the loading to the UASB reactor could not be increased without stressing the system. SPE was working under a tight compliance schedule and could not afford a long start-up lasting several

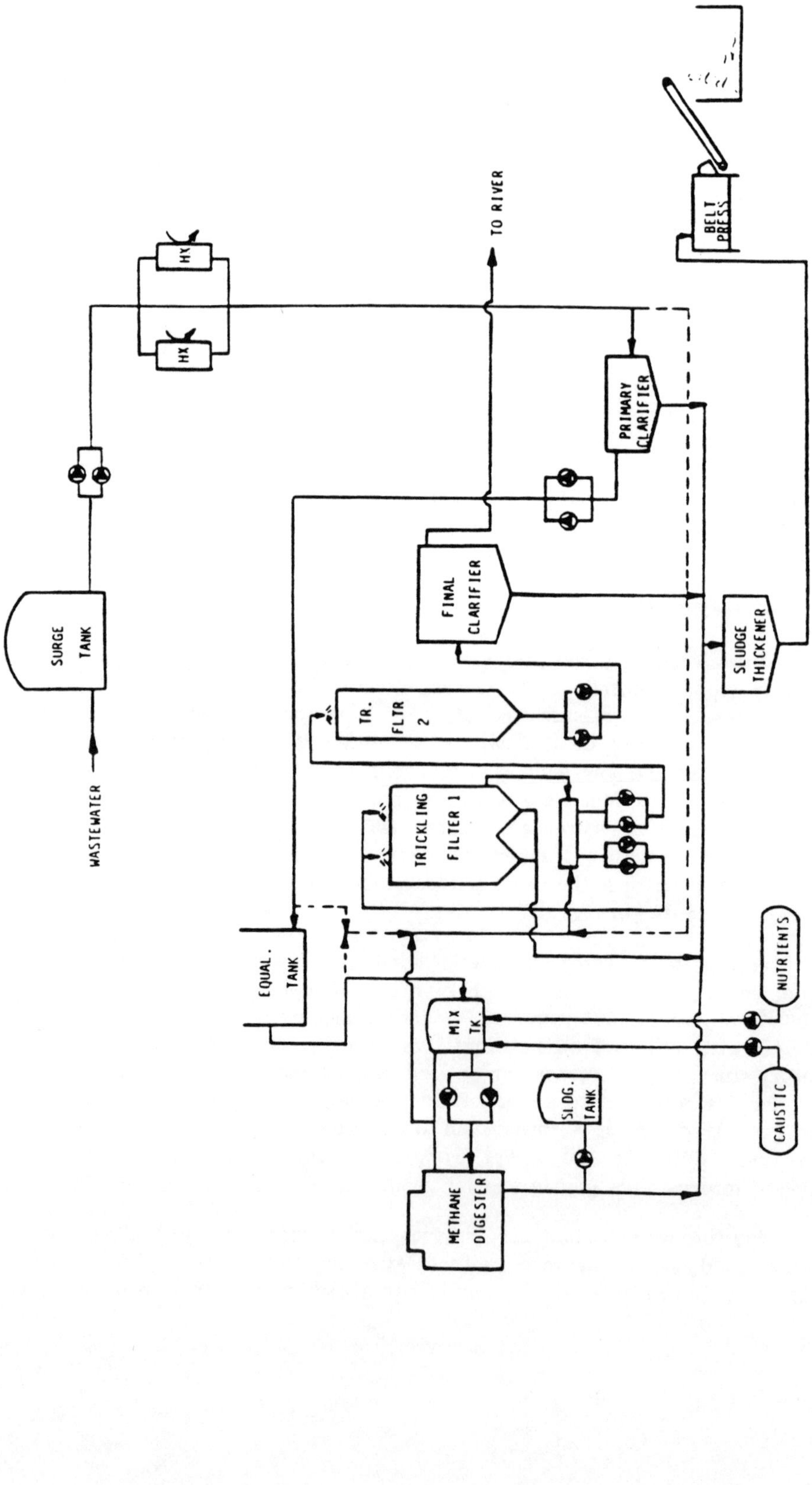

FIGURE 2 Schematic of Treatment Plant After Construction

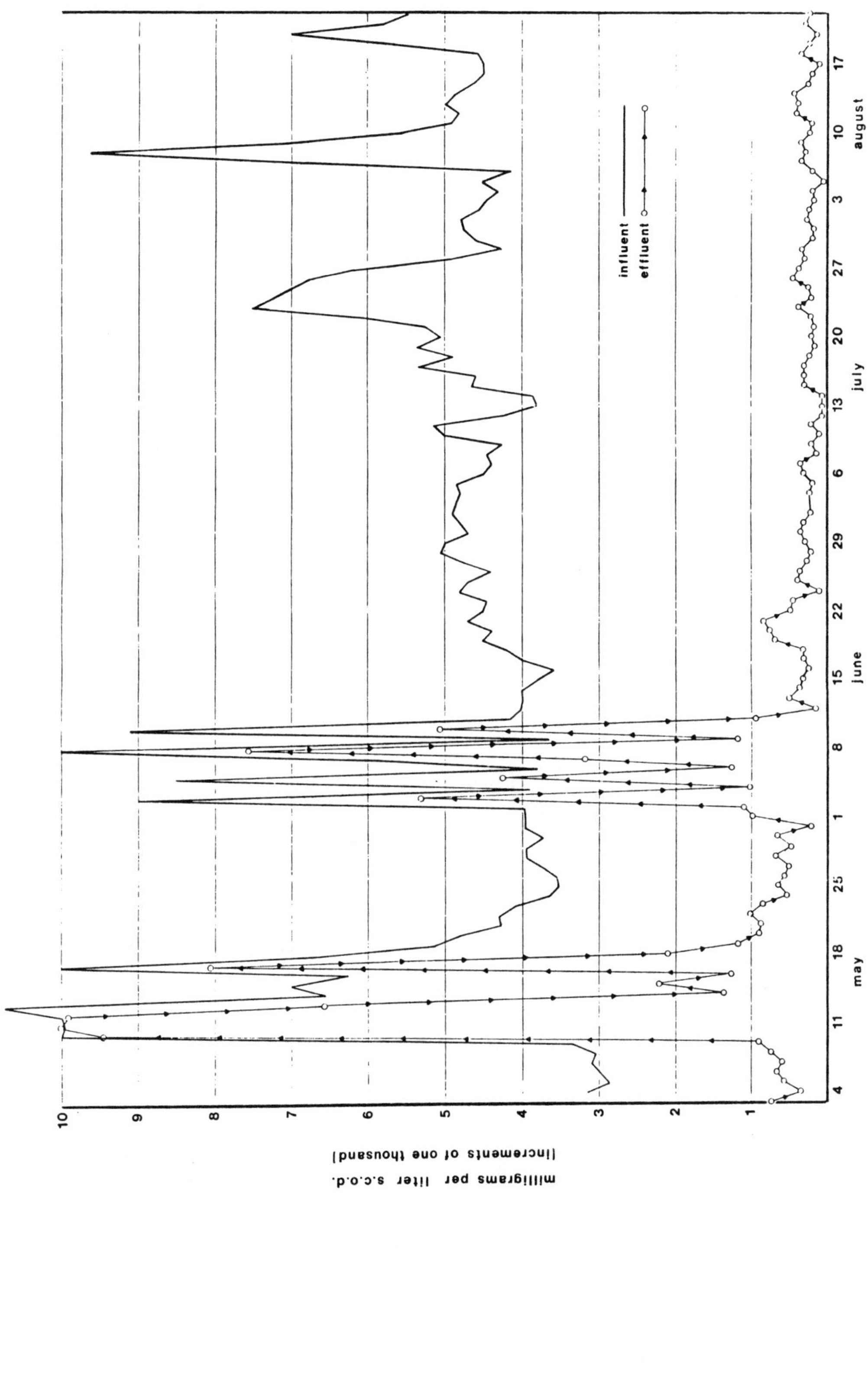

FIGURE 3 Soluble COD Removal from the Anaerobic WTP

months. The anaerobic pretreatment system had to be brought on line as soon as possible. Therefore, it was decided to inoculate the reactor with additional granular seed sludge to accelerate the start-up.

The initial charge of four truck loads of sludge (90 m^3 or 20,000 gls) was completed on June 13. The effect of biomass addition was studied for one week to determine whether the extra biomass inoculation indeed would accelerate start-up. As can be seen from Fig. 3, the process efficiency immediately increased. Since the process efficiency remained satisfactory at slowly increasing loadings to the digester, the decision was made to supplement the digester with sufficient biomass to accept the full design process flow in the shortest possible time. During the next four weeks, an additional 300 m^3 (80,000 gls) of biomass was hauled to SPE. This effectively increased the biomass inventory by about 150% to 25,000 Kg (55,000 lbs) of VSS.

One week into the final sludge hauling operation (June 23 thru 30), the effluent quality remained consistent at steady loading rates. This provided operators with the confidence to accelerate the start up of the anaerobic system. During the second week (July 1 thru 8) the flow was increased to 70 percent of full flow. In the third week loadings were continuously increased and design loading was exceeded on July 11, while treating 95 percent of the available wastewater. In the fourth week the performance test began and the last two truck loads of biomass were added to the reactor.

During the performance test (July 14 thru August 13) the UASB pretreatment system on average reduced the influent soluble COD from 5348 mg/l (range 3850 to 7500) to 281 mg/l (range 50 to 440), a removal efficiency of 95 percent. Flows treated ranged from 136 to 216 m^3/- (600 to 950 gpm). Throughout the performance test the performance remained extremely stable. Volatile fatty acid concentrations - an intermediate product in the biodegradation of organic material into methane - generally remained at non-detectable levels. In the same period volumetric loadings to the anaerobic reactor ranged from 21% below to 117% above the design of 10 Kg COD per cubic meter reactor volume per day. During the peak loading of 21.7 Kg COD/m^3.d the specific sludge activity must have been greater than 1.5 Kg COD converted per Kg VSS per day, which is extremely high and indicative of excellent performance.

An interesting phenomenon occurred shortly after the anaerobic pretreatment unit had started to accept all of the process flow. Most of the biomass on the trickling filter media sloughed off and now is gradually being replaced with a new bio-film. The trickling filters had been operating for two years in a pH range of 6.5 to 7.0 and the micro-organisms in the biofilm were mostly anaerobic or facultatively aerobic because design conditions for the trickling filters were continuously exceeded. When the full flow of wastewater was diverted through the anaerobic system, the pH in the trickling filters went to 8.6 due to CO_2 stripping and the highly alkaline nature of the anaerobic effluent. This abrupt change in pH and the drastic reduction in the food source to the filters resulted in a large loss of filter biomass. Since completing the anaerobic start-up SPE is now in the process of reestablishing growth on the filter media and optimizing the operation and performance of the entire facility.

In conclusion, it may be informative to compare the results of the 6 m^3 (1600 gls) UASB pilot study with the performance of the full scale facility during the one month performance test. Table 2 shows a comparison of the most significant operational parameters.

Determination of total COD is not part of the analytical program to monitor performance of the full-scale plant. Therefore, total COD concentrations were calculated by multiplying the TSS concentrations of influent and effluent by 1.2 and adding the result to the soluble COD concentration. The large difference and the average influent concentration during the pilot study and the average concentration during the recent full scale operation is due to changes in the operation of the ethanol facility. Recent operation has resulted in higher acid levels in the fermenters and subsequently higher SCOD levels in the wastewater. The biogas yield could not be calculated for the full-scale plant, because information on methane content of the biogas was not available. The average biogas flow of the full-scale system during the performance test ranged from 340 to 425 m^3/hr (200 to 250 scfm).

The comparison of pilot and full-scale performance data demonstrates that a study with a 6 m^3 UASB reactor, if carried out properly and conscientiously, will provide good engineering data for a full-scale facility. The scaleup from 6 m^3 to 2050 m^3 can be carried out with confidence.

TREATMENT COST

The total capital cost for the treatment plant's expansion has been used in calculating a treatment cost. This capital cost includes the modifications to the existing plant such as the addition of the primary clarifier and sludge thickener, repair and preparation of the equalization tank and final clarifier, and the acquisition of a heat exchanger, extra pumps, instrumentation and other hardware. For the purposes of cost calculations, it was assumed that the capital cost would be depreciated in five years against 10% compounded interest. Other costs have been based on the most recent numbers that were available.

TABLE 2 Comparison of Pilot Scale and Full Scale System Performance

	Pilot Plant	Full Scale
TCOD in (mg/l)	3627	5655
TCOD our (mg/l)	874	646
Removal (%)	76	89
SCOD in (mg/l)	2889	5348
SCOD out (mg/l)	416	281
Removal (%)	86	95
Volumetric Load (KgTCOD/m^3d)	9.3	12.3
Hydraulic Ret. Time (hrs)	9.4	11.0
Biogas Yield (m^3/Kg CODrem)	0.33	N/A

All costs were added to produce an annual expenditure for operation of the pretreatment facility. This expenditure was divided by the total quantity of COD that the facility was designed to pretreat in a 330 day operating year, i.e. based on a volumetric loading rate of 10 Kg COD per cubic meter reactor volume per day. The result is shown in Table 3 as a cost per unit quantity of COD, including a breakdown into various cost-categories.

In evaluating treatment costs at SPE no credit was taken for the methane being generated. The anaerobic project was justified purely on the basis of required COD reduction to meet NPDES limits and the systems ability to accomplish this under varying conditions with minimal sludge generation. Consideration is now being given to using the methane to offset coal usage in SPE's steam generating unit. Other industries considering the use of anaerobic treatment should not only take credit for methane generation but for the reduction in municipal surcharges which will occur if the facility discharges to a publically owned treatment system. The order to magnitude of these credits is as follows:

- Industrial users of natural gas pay about \$0.26 per therm (100,000 BTU) for interruptible service and \$0.58 per therm for firm service (PSE & G, 1986). Under the design conditions at SPE the lowest quoted price would be equivalent to a savings of \$600 per day or \$0.03/Kg SCOD. The reduction in municipal surcharges was not applicable to SPE, since it is a direct stream discharger. However, the potential impact of this credit can be substantial. Based on data in a survey of the Association of Metropolitan Sewerage Agencies (AMSA, 1983) an average surcharge of \$0.10 per pound of BOD was calculated. The COD to BOD ratio for SPE wastewater is 1.5, resulting in a potential savings in surcharges of \$0.15 per Kg SCOD. This does not take into account other surcharges that may apply such as a TSS and/or volume surcharge.

- In general, municipal treatment costs vary greatly with geographical location. Cost may range from \$0.02 to \$0.03/lb BOD (WPCF, 1986), which for this wastewater is equivalent to \$0.03 to \$0.44/Kg SCOD. Thus, when treatment costs exceed \$0.13/Kg SCOD anaerobic pretreatment can usually be justified from a purely economic standpoint. Please note that caution should be used to extrapolate these numbers to plants of different size. Economy of scale has a definite impact on the capital cost of a system and therefore on the overall pretreatment cost.

TABLE 3 Gross Operating Cost and Cost Breakdown for Pretreatment of Corn-ethanol Wastewater at South Point Ethanol

	Overall Pretreatment Cost
Dollars per Kg SCOD	$0.16
Dollars per lbs SCOD	$0.07
Cost Category	Cost Contribution
Capital cost depreciation	60%
Chemicals	15.5%
Operation and Supervision	13%
Maintenance	6%
Lab and Analytical	4%
Utilities	1.5%

CONCLUSIONS

In summary the following conclusions can be drawn:

1. Significant biomass addition was essential for an accelerated start-up.

2. During the performance test the NASB pretreatment system on average reduced the wastewater SCOD from 5348 mg/l to 281 mg/l, a removal efficiency of 95 percent, at volumetric loadings ranging from 7.9 to 21.7 Kg SCOD/m^3 d.

3. The full-scale facility performed better than the pilot plant.

4. The anaerobic system demonstrated its ability to accept large fluctuations in loading, without a detrimental effect on performance.

5. The cost of operating this pretreatment facility is approximately 13 cents per Kg of SCOD removed.

6. Considering the above anaerobic technology appears to be ideally suited for pretreatment of wastewater generated at an ethanol facility or from similar industries.

REFERENCES

1. J. Lanting and R.L Gross. (1985). "Anaerobic Pretreatment of Corn Ethanol Production Wastewater". Proceedings of the 40th Industrial Waste Conference. Purdue University, 905-914.

2. PSE & G (1986) Personal communication.

3. ASMA (1983). "Municipal Wastewater Treatment Plant Financing Survey Results". Association of Metropolitan Sewerage Agencies, Washington, D.C.

4. WPCF (1986). Personal communication.

MICROBIOLOGICAL ASPECTS OF ANAEROBIC DIGESTION

Ann Wilkie

Department of Microbiology and Cell Science
University of Florida–IFAS
Gainesville, Florida

Emer Colleran

Department of Microbiology
University College
Galway, Ireland

ABSTRACT

This paper reviews aspects of the microbiology of anaerobic digestion and focusses on the importance of interspecies reactions in the operation and control of anaerobic reactors. The potential role of micronutrient supplementation in the selection and optimization of digester populations is discussed. The application of specific methanogenic activity test procedures, phospholipid and archaebacterial lipid analyses and methanogen immunotyping to the characterization of digester biomass is reviewed. Current microbiological approaches to the study of methanogenic biofilm and granule formation are also discussed.

INTRODUCTION

The energy crisis of the seventies greatly stimulated engineering research on anaerobic digestion of industrial wastewaters, both in Europe and in North America, and resulted in the development of a new generation of anaerobic digester designs based on biomass recycle or on biomass retention independent of waste flow. This reduced reactor volume requirements and improved process stability and control, counteracting the early feelings of unreliability associated with anaerobic treatment. These advances stimulated adoption of the technology at full-scale for industrial wastewater treatment, as evidenced (Table 1) by the EEC-commissioned survey of biogas plants in Europe (1) and by the more recent review of biogas developments in North America by Pohland and Harper (2).

Major advances have also been made during the past decade, in our understanding of the microbiology of anaerobic digestion. Greater appreciation of the importance of bacterial interactions is already providing more informed guidelines for anaerobic digester operation and control. This paper reviews aspects of current microbiological research and indicates their potential impact on the continued development of anaerobic digestion technology.

Table 1. Biogas Plants Treating Industrial Wastes in Europe[a] and North America[b]

Reactor Type	Europe		North America	
	Pilot-Scale	Full-Scale	Pilot-Scale	Full-Scale
Anaerobic Contact	2	21	3	2
Anaerobic Filter	9	3	11	5
UASB	4	29	4	5
Fluidised Bed	0	0	6	0
Hybrid UASB/AF	0	0	0	1
Mixed Digester	2	12	4	1
Others	2	5	4	4
Total	19	70	32	18

[a]Data as of 1983 from the EEC biogas survey of Demuynck _et al_. (1) for Belgium, Denmark, France, West Germany, Greece, Ireland, Italy, The Netherlands, United Kingdom and Switzerland. The survey is currently being updated to December 1986 status.
[b]Data collected in 1985 by Pohland and Harper (2).

MICROBIAL INTERACTIONS

Anaerobic reactors present a unique ecosystem in which diverse groups of bacteria catalyze the conversion of complex organic compounds to methane and carbon dioxide in a highly controlled and coordinated fashion (3). Methanogenesis was initially considered to be a two-phase process in which the volatile fatty acid (VFA) and other fermentation end-products of hydrolytic/fermentative bacteria were directly converted to CH_4 and CO_2 by methanogenic species. The multiphase nature of the process was subsequently revealed by the discovery of hydrogen-producing acetogenic bacteria and by a better appreciation of the limited substrate capabilities of methanogens (Fig. 1).

Advances in our knowledge of the microbiology of methanogenesis have highlighted the role of interspecies H_2 transfer in anaerobic digestion. Obligate H_2-producing acetogens (OHPA species) oxidize VFA fermentation products, such as propionate, butyrate, etc., to acetate, CO_2 and H_2 (Fig. 1). Their importance in methanogenesis is evidenced by the finding that propionate and butyrate are the primary precursors of acetate and H_2 in sewage sludge digesters (4). Acetogenic dehydrogenation of VFA is thermodynamically unfavorable when the H_2 partial pressure (P_{H_2}) exceeds 10^{-3} to 10^{-4} atm (5). VFA conversion in digesters is rendered feasible, however, through simultaneous removal of H_2 by H_2-utilizing species, such as the hydrogenotrophic methanogens (5,6). Until recently, OHPA isolates could only be maintained in the laboratory in syntrophic coculture with H_2-utilizing methanogens or sulphate-reducing Desulfovibrio spp. (which provide an alternative electron sink).

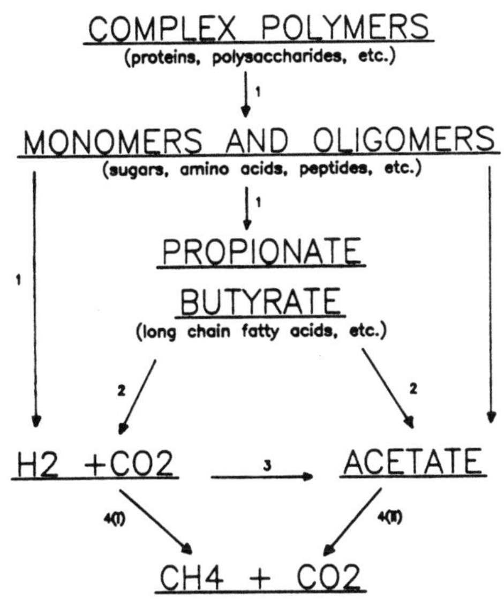

Figure 1. Multiphase Nature of Anaerobic Digestion

1: hydrolytic and non-hydrolytic fermentative bacteria
2: syntrophic acetogens (OHPA)
3: homoacetogens
4: (i) hydrogenotrophic methanogens
(ii) aceticlastic methanogens

To date, only a limited number of OHPA species has been isolated. These include the mesophilic species, Syntrophomonas wolfei (7) and Syntrophobacter wolinii (8), which oxidize butyrate and propionate, respectively. The generation times of these organisms are extremely long, with respective doubling times of 84 and 161 hours for S. wolfei and S. wolinii in methanogenic syntrophic coculture. The recent finding of Beaty and McInerney (9), that S. wolfei can be maintained in pure culture on a crotonate medium, may be expected to accelerate the biochemical study of this key digester group. OHPA species are also known to be involved in the β-oxidation of longer-chain fatty acids (stearate, oleate, etc.) arising from lipid hydrolysis (10) and in the anaerobic degradation of aromatic compounds (11). The doubling time of the benzoate oxidizer isolated by Mountfort et al. (11) is 166 hours in methanogenic coculture. A thermophilic OHPA species which oxidizes butyrate in coculture with a H_2-utilizing methanogen has been isolated by Henson and Smith (12). Interspecies H_2 transfer is also involved in the oxidation of acetate to CH_4 and CO_2 by a thermophilic syntrophic coculture (13).

Maintenance of low P_{H_2} values in digesters is primarily dependent upon the activity of the H_2-utilizing methanogens. Almost all known methanogens convert H_2/CO_2 to CH_4 (14). Mesophilic species exhibit doubling times of the order of 3 or more hours and have K_s values for H_2 in the range 2.5-13 μM (15). They appear to be greatly undersaturated for H_2 in stably operating digesters (16). Methanogenic species which can utilize, as direct methanogenic substrates, the C-2 compound, acetate; C-1 compounds, such as formate, methanol and carbon monoxide, and methylated amines have also been isolated (14). Of these, only acetate is regarded as an important precursor of CH_4 in digesters, being generally recognized as the source of two-thirds or more of the CH_4 evolved, with most of the remainder coming from CO_2 reduction (17). Aceticlastic methanogenesis has been documented for only two methanogenic genera: Methanosarcina (18,19) and Methanothrix (20). With acetate as substrate, doubling times for the mesophilic species, Methanosarcina barkeri and Methanosarcina mazei, are c.24 hours and reported K_s values range from 3 to 5 mM (18,19,21). The filamentous mesophilic aceticlast, Methanothrix soehngenii, displays a considerably lower K_s value for acetate (0.5-0.7 mM) but also exhibits a much slower growth rate, with a doubling time of 3.5-9.0 days (20,22). M. soehngenii is reported to be more abundant in digesters operated at long hydraulic retention times whereas Methanosarcina tend to predominate at shorter retention times (23). A second mesophilic Methanothrix species, designated Methanothrix concilii, has been isolated from sewage sludge (24). Thermophilic Methanothrix spp. have been described by Zinder et al. (25) and by Nozhevnikova and Yagodina (26). A thermophilic Methanosarcina species, Methanosarcina thermophila, has also been isolated and characterized (27,28).

MONITORING AND PROCESS CONTROL

Parameters in current engineering use for routine monitoring of digester performance include organic/hydraulic loading rate, biogas/CH_4 productivity, COD/BOD reduction, alkalinity, pH and VFA concentrations (29). A fall in biogas production rate, an increase in VFA concentration or a change towards acidity are accepted indicators of potential digester failure. Appreciation of the central role of hydrogen in the overall process, however, suggests that these changes may be regarded as the result rather than the cause of digester imbalance. Monitoring of the partial pressure of H_2 is being considered as a potentially sensitive means of early detection of fermentation failure (5). Studies are currently in progress to develop reliable methods for estimation of H_2 levels in digester gases and mixed liquors (30) and to correlate fluctuations in H_2 concentration with changes in feed composition, loading rate and other process variables, for both soluble and particulate feedstocks (31).

Propionic acid may also be used as an indicator of digester balance. The P_{H_2} range in which acetogenic dehydrogenation of propionate and hydrogenotrophic generation of CH_4 can both proceed exergonically is extremely narrow (5,6). Any digester imbalance which results in H_2 partial pressures outside that range should, therefore, cause an immediate increase in propionate. This could be used to signal corrective measures before the microbial community becomes more drastically disturbed by acid accumulation and the consequent fall in pH.

Recent studies on the digestion of complex organic wastes have suggested that other process intermediates may also be used as indicators of the efficiency of coupling between the hydrolytic/acidogenic and the methanogenic phases of the overall process. Phenylacetic acid (PAA) is an intermediate of the degradation of aromatic amino acids and lignocellulosic materials. The conversion of PAA to CH_4 is catalyzed by the syntrophic activity of the OHPA species, Syntrophus buswellii, in association with hydrogenotrophic and aceticlastic methanogens (11). Iannotti et al. (32) reported that effluent PAA levels in digesters treating swine manure appeared to provide a more sensitive indicator of digester stress than volatile acids. Although the data reported is preliminary in nature, it implies that key process intermediates (other than VFA) may be used to monitor digester balance in applications where the feed is rich in lipids or more refractory organic compounds, such as aromatic or halogenated aliphatic compounds, lignocellulose, etc.

EFFECTIVE BIOMASS MEASUREMENT

The assay procedure most commonly used in engineering practice for seed sludge quantification and biomass measurement in operational digesters is the determination of volatile suspended solids (VSS). These analyses provide crude estimates of effective microbial biomass since they also measure non-microbial organic material contributed by wastewaters containing insoluble organics (e.g. animal manures, crop residues, starch wastewaters, etc.). Furthermore, these methods give no indication of the population density or functional activities of different trophic groups in seed sludges or in digester biomass. Direct enumeration techniques are not a viable alternative since they are tedious and time-consuming; require access to anaerobic culture facilities and may yield misleading data, as species of importance in digesters may not necessarily flourish under the diagnostic enrichment/isolation conditions used, and vice versa (33).

Since methanogenic bacteria exhibit a unique biochemistry, the possibility of quantifying the methanogenic biomass in digesters by measurement of unique methanogen components has been investigated by a number of researchers. Coenzyme F_{420} - a deazaflavin analogue which plays a key metabolic role in methanogens as an electron carrier - has received most attention. It is present in all methanogens tested to date and does not occur in other bacteria, with the exception of Streptomyces griseus and Anacystis nidulans. Although simple, reliable procedures are now available for F_{420} measurement (34,35), wide variation in individual methanogen F_{420} levels (Table 2) complicates interpretation of measured sludge F_{420} concentrations (35,36). Observations by van Bruggen et al. (37), that the coccoid form of Methanosarcina mazei exhibits a weak F_{420} fluorescence compared to the strong fluorescence of the sarcinal aggregate, suggest that F_{420} levels may also vary with morphological form, thus further complicating interpretation of sludge F_{420} data.

Table 2. Coenzyme F_{420} Content of Mesophilic Methanogens[a]

Organism	F_{420} (nMole.gVSS^{-1})
H_2-utilizing methanogens	
Methanobacterium bryantii	1,400-2,400
Methanobrevibacter ruminantium	40
Methanobrevibacter arboriphilus	1,800
Methanospirillum hungatei	1,900
Formate-utilizing methanogens	
Methanobacterium formicicum	3,000-3,600
Acetate-utilizing methanogens	
Methanosarcina acetivorans	27
Methanosarcina barkeri, 277	9[b]
Methanosarcina barkeri, MS	19[c]
Methanothrix soehngenii	330

[a]Data adapted from Dolfing and Mulder (38).
[b]Expressed as nMole.g protein^{-1} (39).
[c]Expressed as nMole.g (wet wt) cells^{-1} (40).

Correlation of F_{420} concentrations with potential methanogenic activity for individual substrates is also extremely difficult, since the F_{420} levels maintained by acetate-utilizing methanogens are much lower than those of the H_2/CO_2 or formate-utilizing species (Table 2). A recent study by Dolfing and Mulder (38), using granular sludges cultivated on various direct and indirect methanogenic substrates (acetate, ethanol, propionate, etc.), revealed a good correlation only with formate as substrate. Measurement of sludge F_{420} levels may, however, be useful in determining the vertical distribution of methanogenic bacteria in upflow or downflow fixed-bed reactors or the distribution of methanogens between the biofilm and mixed liquor in fixed-bed and hybrid reactors (41).

The distinctive archaebacterial membrane lipids present in all methanogens contain unique phytanyl glycerol ethers which can be extracted, purified and quantified by HPLC procedures (42). Consequently, measurement of the phytanyl glycerol ether content of anaerobic sludges has potential as a technique for estimating methanogenic biomass content. For non-methanogenic trophic groups, measurement of membrane phospholipid esters has been shown to provide a good estimate of microbial biomass and to correlate well with extractable ATP measurements (43). In addition, analysis of individual phospholipid fatty acids may permit more detailed characterization of microbial community structure since some fatty acids are unique to individual bacteria, or to groups of bacteria, and may function as molecular signatures for these species (44). Measurement of both phytanyl glycerol ethers and membrane phospholipids and identification of individual phospholipid fatty acids, in anaerobic sludges, may prove useful in determining the relative contents of methanogenic and non-methanogenic biomass; in monitoring changes in the ratio of the two populations and in assessing individual species distribution in response to changes in loading rate, presence of toxicants, etc. Lipid analyses may also be used to distinguish between microbial and non-microbial VSS in digesters treating plant biomass or animal waste feedstocks (43).

BIOMASS ACTIVITY MEASUREMENTS

Detailed knowledge of the activities of different trophic groups in seed sludges, mixed liquors, biofilm and granule samples should assist in defining start-up conditions and assessing maximum loading capacities for operational reactors. Earlier methods, which were confined to estimating methanogenic activity levels with acetate as substrate, required large amounts of sludge and were both time-consuming and error-prone (35). Routine test procedures have now been developed, based on gas chromatography and pressure-lock syringe sampling (45) or pressure-transducer monitoring (35) of the headspace in test vials. These methods are sensitive, rapid and reliable, and can be utilized to determine the potential methanogenic activities of digester sludges against the full range of direct and indirect methanogenic substrates. Activity test procedures may also be used to assess the inhibitory effect of toxicants on individual methanogenic and OHPA populations in crude sludges; to study the effect of changes in process parameters on the distribution and activity of these populations and to characterize the microbiological composition of biofilms and granules developed in reactors under different loading rates and with varying feed compositions.

ANAEROBIC BIOFILM FORMATION

Start-up and long-term operation of fixed-bed reactors, particularly if operated in downflow mode with modular or channelled support materials, is dependent upon the initial development of a competent biofilm on the support surface and upon its continued maintenance in a form which ensures efficient contact and reaction of the influent substrate with the retained flora. It is important, therefore, to determine the influence of the support material on the attachment of different trophic groups and of individual species within these groups. Such studies have already been initiated (46,47) and may be expected to provide a more valid basis for choice of support material type and packing arrangement. For example, efficient biofilm retention of slow-growing organisms, such as the OHPA bacteria or the aceticlast, Methanothrix soehngenii, could accelerate start-up and ensure more stable operation and more efficient COD removal at possibly higher loading rates. Ongoing studies have shown that Methanosarcina mazei MC3 does not attach to a wide range of plastic and clay supports at the initial adhesion stage of biofilm formation (48). Methanosarcina spp. have, however, been reported to play a dominant role in anaerobic biofilms during the start-up phase (35,49). The possibility that the presence of Methanosarcina spp. results from passive entrapment, rather than active involvement in biofilm formation, merits further study. Studies of this kind should assist in the definition of start-up regimes and could be economically beneficial if they result in significantly shorter start-up times for fixed-film reactors.

SLUDGE GRANULATION

Studies on mesophilic granule formation have shown that varied granular forms may be cultivated on different wastewaters and under different start-up conditions (50,51). Granules developed on mainly VFA feeds tend to be of the "filamentous" type, up to 5 mm in size and mechanically fragile. They are reported to contain inert carrier material and to be dominated by a highly filamentous form of Methanothrix, presumed to be M. soehngenii. More robust, "rod"-type granules develop on sugar beet or potato processing wastewaters. These are generally smaller (up to 3 mm in size), contain no detectable inert carrier and are again dominated by M. soehngenii-like species, but in a much shorter chain-length form (50,51). Granules containing Methanosarcina spp. as the dominant aceticlast can also be cultivated under high VFA concentrations and low pH (50,52). However, Lettinga et al. (52) consider their development in reactors to be undesirable since they exhibit poor activity at low acetate concentrations and are prone to washout because of their small size (< 0.5 mm). A fourth type of granule has been reported from a full-scale plant treating maize starch production wastewater which contained an exceptionally high Ca^{++} content (> 700 $mg.L^{-1}$). These "spiky" granules (c. 1 mm in length) were dominated by filamentous organisms and contained up to 60% by weight of $CaCO_3$ (50,51).

At concentrations below 200-300 ppm, Ca^{++} has been shown to exert a positive effect on granule formation (53). This has been attributed to the chelating effect of Ca^{++} bridges and to the increased density arising from $CaCO_3$ precipitation (54). Physicochemical studies by Mahoney et al. (55) also provide further evidence of the importance of divalent cation bridging in the initial stages of granulation. Higher concentrations of Ca^{++} adversely affect granule formation due to excessive precipitation of $CaCO_3$ and $(Ca)_3(PO_4)_2$ (51). This causes scaling of the outer surfaces of pre-formed granules, increasing granule density and decreasing sludge activity. The high sludge concentrations which result at the base of the reactor make efficient contacting of the wastewater with the sludge impossible (52).

Formation of thermophilic granules has also been investigated. Wiegant et al. (56) initially failed to obtain granulation during a 300 day study with pure VFA mixtures. Granulation occurred rapidly, however, with glucose or sucrose feedstocks. The granules obtained differed from the common mesophilic forms, being smaller in size (c. 1.3 mm) and consisting mainly of thermophilic Methanosarcina spp. (56). Successful formation of very small thermophilic granules (c. 0.2 mm), on a mixture of acetic, propionic and lactic acids, has recently been reported by Endo and Watanabe (57). Much larger aggregates (c. 3.0 mm) have been obtained by Bochem et al. (58) in chemostat studies of acetate enrichments. These granules consisted of densely-packed Methanosarcina clusters surrounding a more loosely-packed central area which contained at least two non-methanogenic species.

Adaptation of granular sludge, cultivated initially on one type of industrial wastewater, to a different waste stream has not always been successful at laboratory or at full-scale. Early studies by Lettinga et al. (59) showed that granules with high specific activity and excellent settling properties form readily on sugar beet processing wastewater. Transfer of granules, developed on sugar beet or corn starch wastewaters, to a sucrose feedstock resulted in the development of voluminous layers of bacterial biomass and gelatinous exopolymers on the granule surfaces, causing a severe decrease in sludge settleability and ultimate washout from the reactor (60). These authors concluded that a sudden change in feed composition cannot always be accommodated by granular sludge developed initially on mainly acidified effluents. Granules with excellent settling properties could, however, be formed ab initio with sucrose-containing feedstocks (60).

Similarly, initiation of granulation with more recalcitrant wastewaters has not always been successful, as shown by pilot-scale studies on sludge conditioning liquor in Canada (61). Granulation was not obtained during 360 days of operation, despite repeated re-seeding, modification of the gas-liquid-solids separator and eventual operation in anaerobic contact mode with effluent solids settlement and recycle. Change to a brewery wastewater, however, resulted in granulation within six weeks (61).

Although granulation occurs with a variety of wastewaters, further studies are required in order to ascertain the roles of individual bacterial groups or species and of insoluble inorganic materials (e.g. calcium) in effective granule formation. Identification and study of the Methanothrix species which dominate mesophilic "filamentous" or "rod"-type granules is of particular interest, in view of reported differences in doubling times and K_s values for acetate between M. soehngenii and M. concilii (20,24). The possible involvement of carbohydrate decomposing bacteria, as suggested by Dolfing and Bloemen (62), also merits investigation.

MICRONUTRIENT EFFECTS

An increasing number of digester studies indicates considerable stimulation of methanogenesis by supplementation with trace elements (63,64,65).

Bioavailability of minerals to digester microorganisms is difficult to measure and what appears to be a sufficient concentration of individual micronutrients in a feedstock may not equate with adequate bioavailability (64).

Recent studies by Speece et al. (65), on methane production from acetate in mixed reactors receiving unlimited substrate and operated at 10 or 20 day HRT/SRT, highlighted interesting effects of nutrient supplementation. The enrichment culture, which had been maintained in the laboratory for eight years on a synthetic medium containing a variety of inorganic nutrients, appeared to be dominated by Methanothrix-like rods. Additional supplementation with 13 $mg.L^{-1}$ Fe^{++} above the basal level was shown to change the microbiological dominance of the enrichment within 10 days to a Methanosarcina mazei-like organism and to increase the maximum acetate utilization rate from 3-4 to 30-40 g $acetate.L^{-1}.d^{-1}$. This change was inhibited if either phosphate or sulfide anions were added with the extra Fe^{++}, indicating that bioavailability of the iron was essential for the population change (65). The higher acetate utilization rates were sustainable for only a short period (c. 7 days), during which the Methanosarcina-like cells were shown to become associated in large aggregate form. The acetate utilization rate then dropped abruptly to c. 5 $g.L^{-1}.d^{-1}$. This was shown to correlate with a break-up of the aggregates to individual coccoid cells. Maintenance of Ca^{++} and Mg^{++} levels above 200 $mg.L^{-1}$ prevented de-aggregation and maintained high acetate utilization rates for prolonged periods (65).

Pure culture studies by Scherer et al. (66) have yielded conflicting results. Omission of Ca^{++} from the growth medium resulted in flocculant growth of Methanosarcina barkeri strain Fusaro (flocs of 1 mm size), whereas addition of 2 mM $CaCl_2$ caused an immediate change to a dispersed growth form. By contrast, M. barkeri strain Julich grew in large aggregate form (c. 3 mm) in the presence of Ca^{++}. Ability to aggregate appears to be a strain-specific property of strain Julich and to be associated with a significantly higher polysaccharide content, rich in glucuronic acid residues (66).

These, and other studies, suggest that the bioavailability of a variety of trace elements, vitamins and other nutrients may have dramatic effects on the microbial composition of digester biomass. A better understanding of the microbial requirements for, and the bioavailability of, trace elements and other nutrients in anaerobic reactors may permit greater control of both the composition and the activity of the microbial populations involved.

BIOMASS CHARACTERIZATION

The immunological procedures developed by Conway de Macario and Macario (67,68) appear to offer most promise for the rapid identification of methanogenic species active in anaerobic digesters. These authors have assembled a comprehensive bank of antisera against 30 reference methanogens available in pure culture (69) and have developed a series of calibrated polyclonal antibody probes from the antisera bank. Testing of individual methanogens against all of the available probes has provided a sequence of data, distinctive for each methanogen and referred to as its antigenic fingerprint. Arrangement of the fingerprint data in a matrix formed by the phylogenetic organization of methanogens clearly showed that antigenic relatedness coincides with phylogenetic kinship (67). Arrangement of the data in this way permitted selection of representative methanogens for development of panels of monoclonal antibodies using mouse hybridoma techniques (68). Many of the monoclonal antibodies studied to date are monospecific, reacting exclusively with a single antigenic determinant on a particular methanogen immunotype (69).

The availability of polyclonal and monoclonal antibody probes will permit precise identification and quantification of individual methanogens - not only at the species or strain level but at the immunotype level - in microlitre samples of digester mixed liquor, biofilm and granule preparations. In particular, this will permit identification of immunotypes active under start-up conditions; assessment of immunotype changes in response to duration of operation, alteration of HRT or increase in loading rate; correlation of immunotypes with wastewater composition; identification of predominant immunotypes in granules and in biofilms developed on different support materials, etc. (70). Ultimately, such information should prove invaluable for definition of seeding, start-up and operational regimes and for optimization of reactor design with respect to individual wastewater types.

CONCLUSION

Developments in reactor design and operation have established anaerobic digestion as an accepted process for industrial wastewater treatment. Parallel advances in our understanding of the complex microbiology of the process are providing new insights into microbial interactions and into factors governing the dominance, activity and maintenance of individual species in digester mixed liquors, biofilms and granules. This knowledge will contribute significantly to improved design, start-up, process control and operation of anaerobic digesters, with corresponding effects on the efficiency, stability and cost effectiveness of full-scale anaerobic waste treatment applications.

Use of monoclonal antibody probes may be expected, in the long term, to identify the methanogen immunotypes most promising for genetic engineering manipulation (70). The discovery of methanogen plasmids (71), the identification of a phage specific for a *Methanobrevibacter* species (72) and the more recent isolation and characterization of a *Methanothrix* phage (73) have increased the possibility of developing genetic exchange systems for methanogens. In addition, cloning studies have been successful in obtaining expression of functional methanogen gene products in *Escherichia coli* and *Bacillus subtilis* (74,75). These developments suggest that future microbiological research will continue to complement process and engineering studies, expanding the application of anaerobic digestion technology in the areas of wastewater treatment and renewable energy production.

REFERENCES

1. Demuynck, M., Nyns, E.J. and Palz, W. (1984). Biogas Plants in Europe. D. Reidel Publishing Company, Dordrecht, Holland.
2. Pohland, F.G. and Harper, S.R. (1985). In: Anaerobic Digestion 1985, p.41-81. The China State Biogas Association, Beijing, China.
3. McInerney, M.J. and Bryant, M.P. (1981). In: Fuel Gas Production from Biomass, Vol.1, p.19-46. D.L. Wise (ed.), CRC Press, Inc., Boca Raton, Florida.
4. Smith, P.H. (1980). Report No. EPA-600/2-80-093. U.S. Environmental Protection Agency, Cincinnati, Ohio.
5. Wolin, M.J. and Miller, T.L. (1982). ASM News **48**:561-565.
6. Boone, D.R. (1985). In: Biotechnological Advances in Processing Municipal Wastes for Fuels and Chemicals, p.85-96. A.A. Antonopoulos (ed.), ANL/CNSV-TM-167, Argonne National Laboratory, Argonne, Illinois.
7. McInerney, M.J., Bryant, M.P., Hespell, R.B. and Costerton, J.W. (1981). Appl. Environ. Microbiol. **41**:1029-1039.
8. Boone, D.R. and Bryant,M.P. (1980). Appl. Environ. Microbiol.**40**:626-632.
9. Beaty, P.S. and McInerney, M.J. (1986). Abstr. Annu. Meet. Am. Soc. Microbiol., I-132, p.187.
10. Roy, F., Albagnac, G. and Samain, E. (1985). Appl. Environ. Microbiol. **49**:702-705.
11. Mountfort, D.O., Brulla, W.J., Krumholz, L.R. and Bryant, M.P. (1984). Int. J. Syst. Bacteriol. **34**:216-217.
12. Henson, J.M. and Smith,P.H. (1985). Appl.Environ.Microbiol.**49**:1461-1466.
13. Zinder, S.H. and Koch, M. (1984). Arch. Microbiol. **138**:263-272.
14. Balch, W.E. and Wolfe, R.S. (1979). J. Bacteriol. **137**:264-273.
15. Robinson, J.A. and Tiedje, J.M. (1984). Arch. Microbiol. **137**:26-32.
16. Kaspar, H.F. and Wuhrmann, K. (1978). Appl. Environ. Microbiol. **36**:1-7.
17. Smith, P.H. and Mah, R.A. (1966). Appl. Microbiol. **14**:368-371.
18. Smith, M.R., Zinder, S.H. and Mah,R.A. (1980). Process Biochem.**15**:34-39.
19. Mah, R.A. (1980). Curr. Microbiol. **3**:321-326.
20. Huser, B.A., Wuhrmann, K. and Zehnder, A.J.B. (1982). Arch. Microbiol. **132**:1-9.
21. Zinder, S.H. (1984). ASM News **50**:294-298.
22. Krzycki, J.A. and Zeikus, J.G. (1985). In: Biotechnological Advances in Processing Municipal Wastes for Fuels and Chemicals, p.15-34. A.A. Antonopoulos (ed.), ANL/CNSV-TM-167, Argonne National Laboratory, Argonne, Illinois.
23. Mah, R.A. (1983). In: Third International Symposium on Anaerobic Digestion - Proceedings, p.13-22. The Third International Symposium on Anaerobic Digestion, 99 Erie Street, Cambridge, Massachusetts.
24. Patel, G.B. (1984). Can. J. Microbiol. **30**:1383-1396.
25. Zinder, S.H., Cardwell, S.C., Anguish, T., Lee, M. and Koch, M. (1984). Appl. Environ. Microbiol. **47**:796-807.
26. Nozhevnikova, A.N. and Yagodina, T.G. (1982). Microbiology **51**:534-541.
27. Zinder, S.H. and Mah, R.A. (1979). Appl. Environ. Microbiol.**38**:996-1008.
28. Zinder, S.H., Sowers, K.R. and Ferry, J.G. (1985). Int.J.Syst.Bacteriol. **35**:522-523.
29. Archer, D.B. (1983). Enzyme Microb. Technol. **5**:162-170.
30. Scott, R.I., Williams, T.N., Whitmore, T.N. and Lloyd, D. (1983). Eur. J. Appl. Microbiol. Biotechnol. **18**:236-241.

31. Smith, P.H. (1986). Personal communication. University of Florida, Gainesville, Florida.
32. Iannotti, E.L., Mueller, R.E., Sievers, D.M., Georgacakis, D.G. and Gerhardt, K.O. (1986). J. Ind. Microbiol. 1:57-61.
33. Dolfing, J., Griffioen, A., van Neerven, A.R.W. and Zevenhuizen, L.P.T.M. (1985). Can. J. Microbiol. 31:744-750.
34. Hutschemackers, J., Delafontaine, M., Naveau, H.P. and Nyns, E.-J. (1982). Biomass 2:115-125.
35. Reynolds, P.J. (1986). Ph.D. Thesis. National University of Ireland, Dublin, Ireland.
36. Zabranska, J., Schneiderova, K. and Dohanyos, M. (1985). Biotechnol. Lett. 7:547-552.
37. van Bruggen, J.J.A., Goosen, N.K. and Stumm, C. (1984). Antonie van Leeuwenhoek 50:80-81.
38. Dolfing, J. and Mulder, J-W. (1985). Appl. Environ. Microbiol. 49:1142-1145.
39. Baresi, L. and Wolfe, R.S. (1981). Appl. Environ. Microbiol. 41:388-391.
40. Eirich, L.D., Vogels, G.D. and Wolfe,R.S. (1979). J.Bacteriol.140:20-27.
41. Reynolds, P.J. and Colleran, E. (1986). In: Anaerobic Treatment A Grown-Up Technology, p.515-531. Industrial Presentations (Europe) B.V., Schiedam, The Netherlands.
42. Tornabene, T.G. and Langworthy, T.A. (1979). Science 203:51-53.
43. Henson, J.M., Smith, P.H. and White, D.C. (1985). Appl. Environ. Microbiol. 50:1428-1433.
44. White, D.C. (1983). In: Microbes in Their Natural Environments, p.37-66. J.H. Slater, R. Whittenbury and J.W.T. Wimpenny (ed.), Cambridge University Press, Cambridge, England.
45. Dolfing, J. and Bloemen, W.G.B.M. (1985). J. Microbiol. Methods 4:1-12.
46. Verrier, D. and Albagnac,G. (1985). In: Energy from Biomass, p.537-541. W. Palz, J. Coombs and D.O. Hall (ed.), Elsevier Applied Science Publishers, London.
47. Switzenbaum, M.S. (1986). Report No. Env.E. 90-86-1. Department of Civil Engineering, University of Massachusetts, Amherst, Massachusetts.
48. Verrier, D. (1986). Personal communication. INRA, Villeneuve d'Ascq, France.
49. Robinson, R.W., Akin, D.E., Nordstedt, R.A., Thomas, M.V. and Aldrich, H.C. (1984). Appl. Environ. Microbiol. 48:127-136.
50. Hulshoff Pol, L.W., de Zeeuw, W., Dolfing, J. and Lettinga, G. (1983). In: Anaerobic Waste Water Treatment, p.40-43. W.J. van den Brink (ed.), TNO, The Hague, The Netherlands.
51. Lettinga, G., Hulshoff Pol, L.W., Hobma, S.W., Grin, P., de Jong, P., Roersma, R. and IJspeert, P. (1983). In: Anaerobic Waste Water Treatment, p.411-429. W.J. van den Brink (ed.), TNO, The Hague, The Netherlands.
52. Lettinga, G., de Zeeuw, W., Hulshoff Pol, L., Wiegant, W. and Rinzema,A. (1985). In: Anaerobic Digestion 1985, p.279-301. The China State Biogas Association, Beijing, China.
53. Hulshoff Pol, L.W., de Zeeuw, W.J., Velzeboer, C.T.M. and Lettinga, G. (1983). Wat. Sci. Tech. 15(8/9):291-304.
54. Brunetti, A., Boari, G., Passino, R. and Rozzi, A. (1983). In: Anaerobic Waste Water Treatment, p.317-334. W.J. van den Brink (ed.), TNO, The Hague, The Netherlands.

55. Mahoney, E.M., Varangu, L.K., Cairns, W.L. and Kosaric, N. Presented at the 57th WPCF Conference, New Orleans, Louisiana, October 1-4, 1984.
56. Wiegant, W.M., Claassen, J.A., Borghans, A.J.M.L. and Lettinga, G. (1983). In: Anaerobic Waste Water Treatment, p.392-410. W.J. van den Brink (ed.), TNO, The Hague, The Netherlands.
57. Endo, G. and Watanabe, A. (1986). Abstr. Annu. Meet. Am. Soc. Microbiol., I-125, p.185.
58. Bochem, H.P., Schoberth, S.M., Sprey, B. and Wengler, P. (1982). Can. J. Microbiol. 28:500-510.
59. Lettinga, G., van Velsen, A.F.M., Hobma, S.W., de Zeeuw, W. and Klapwijk, A. (1980). Biotechnol. Bioeng. XXII:699-734.
60. Hulshoff Pol, L.W., Dolfing, J., van Straten, K., de Zeeuw, W.J. and Lettinga, G. (1984). In: Current Perspectives in Microbial Ecology, p.636-642. M.J. Klug and C.A. Reddy (ed.), American Society for Microbiology, Washington, D.C.
61. Hall, E.R. and Jovanovic, M. (1983). In: Proceedings of the 37th Industrial Waste Conference, p.719-728. Ann Arbor Science Publishers, Ann Arbor, Michigan.
62. Dolfing, J. and Bloemen, W. (1984). Antonie van Leeuwenhoek 50:86.
63. Murray, W.D. and van den Berg, L. (1981). Appl. Environ. Microbiol. 42: 502-505.
64. Wilkie, A., Goto, M., Bordeaux, F.M. and Smith, P.H. (1986). Biomass 11: 135-146.
65. Speece, R.E., Parkin, G.F., Bhattacharya, S. and Takashima, M. (1986). In: Anaerobic Treatment A Grown-Up Technology, p.175-188. Industrial Presentations (Europe) B.V., Schiedam, The Netherlands.
66. Scherer, P.A., Bochem, H.-P., Davis, J.D. and White, D.C. (1986). Can. J. Microbiol. 32:137-144.
67. Macario, A.J.L. and Conway de Macario, E. (1985). In: Monoclonal Antibodies against Bacteria, Vol.2, p.213-247. A.J.L. Macario and E. Conway de Macario (ed.), Academic Press, Inc., Orlando, Florida.
68. Conway de Macario, E. and Macario, A.J.L. (1985). In: Anaerobic Digestion 1985, p.505-514. The China State Biogas Association, Beijing, China.
69. Conway de Macario, E. and Macario, A.J.L. (1986). System. Appl. Microbiol. 7:320-324.
70. Macario, A.J.L. and Conway de Macario, E. (1987). In: Energy from Biomass and Wastes X, p.1009-1020. Institute of Gas Technology, Chicago, Illinois/Elsevier Applied Science Publishers, London.
71. Thomm, M., Altenbuchner, J. and Stetter, K.O. (1983). J. Bacteriol. 153: 1060-1062.
72. Baresi, L. and Bertani, G. (1984). Abstr. Annu. Meet. Am. Soc. Microbiol., I-74, p.133.
73. Roustan, J.L., Touzel, J.P., Prensier, G., Dubourguier, H.C. and Albagnac, G. (1986). In: Biology of Anaerobic Bacteria, p.200-205. H.C. Dubourguier, G. Albagnac, J. Montreuil, C. Romond, P. Sautiere and J. Guillaume (ed.), Elsevier Science Publishers B.V., Amsterdam, The Netherlands.
74. Wood, A.G., Redborg, A.H., Cue, D.R., Whitman, W.B. and Konisky, J. (1983). J. Bacteriol. 156:19-29.
75. Morris, C.J. and Reeve, J.N. (1985). In: Biotechnological Advances in Processing Municipal Wastes for Fuels and Chemicals, p.175-186. A.A. Antonopoulos (ed.), ANL/CNSV-TM-167, Argonne National Laboratory, Argonne, Illinois.

REACTOR DESIGN CONSIDERATIONS AND EXPERIENCES WITH VARIOUS WASTEWATERS

Ton J.M. Jans and Gijs De Man

Grontmig Consulting Engineers
De Holle Bilt 22, De Bilt, The Netherlands

ABSTRACT

During the last decade a great number of high-rate anaerobic treatment plants were built and put into operation for the (pre)treatment of mainly industrial wastewaters. The breakthrough in anaerobic technology for relatively diluted wastes (water content over 95 %) resulted from the development and successful application of new, simple and relatively inexpensive anaerobic treatment processes.
This paper discusses a selection of reactor design considerations of the upflow anaerobic sludge blanket process (UASB). This overview is based on experience with several full-scale plants presently in operation, as well as on laboratory and pilot-plant research. Finally, some project descriptions of operational reactors are presented. These descriptions discuss specific experiences encountered during start-up and operation of the reactors.

INTRODUCTION

Although anaerobic digesters have traditionally been used for many decades in the stabilization of sewage sludge, their successful and economic employment for the treatment of liquid wastes is only a recent phenomenon, arising from the development of new reactor designs. These concepts have in common that they are capable of retaining a much higher biomass concentration than traditional digesters. This is made possible by making the sludge retention independent of the influent retention time.

The technological approaches to allow this condition of independent sludge retention can be divided into:

- attached biomass concepts (filters and fluidized systems);
- non-attached concepts (sludge blanket reactors and the contact process with sludge recycling).

It is difficult to evaluate the advantages and disadvantages of each system in relation to other concepts, as generalizations are not usually valuable in practice. Considerations such as:

- purification rates,
- loading rates,
- investment cost,
- energy balance,
- space requirements,
- operational costs,
- specific long-term experience with certain wastewaters,

are all of importance, but they will be valued differently from industry to industry.

However, as can be seen in the overview of operational high-rate anaerobic reactors (Table 1), the UASB process has been widely applied and has proved to be suitable for the treatment of a wide variety of industrial wastewaters both in origin and concentration.

Table 1. UASB reactors installed and commissioned (July 1986, total = 82)

Substrate	Number	Design Load kg COD/(m^3.day)
Sugar	19	7.7-18
Potato processing	15	4 -10.7
Starch (potato, maize, wheat)	9	7 -13
Brewery	4	6.8-14.5
Alcohol	13	9.5-15
Divers (yeast, slaughterhouse)	22	3 -29

Grontmij Consulting Engineers have been involved in laboratory and pilot studies for specific wastewaters and in design and start-up of full-scale anaerobic treatment plants for over 10 years. Our company operates in close co-operation with the Agricultural University in Wageningen, which played an important role in the development of this technology, under the leadership of Dr Lettinga.

THE UASB REACTOR

The main elements of the reactor (Figure 1) are the influent distribution system in the bottom of the reactor, and a three-phase separator (gas, solids, liquid) at the top. The influent is distributed over the bottom and mixed with the anaerobic sludge bed by the influent distribution system. The organic compounds are removed from the wastewater as it rises to the top of the reactor and are converted mainly into biogas and some new sludge. The biogas produced during this process increases the contact between sludge and feed. The anaerobic sludge and the biogas are separated from the treated effluent in the three-phase separator. With the absence of mechanical mixing and sludge recirculation, and because of the favourable physical and chemical conditions, a granulation process takes place in the reactor. This results in the formation of an anaerobic sludge with excellent settling characteristics.

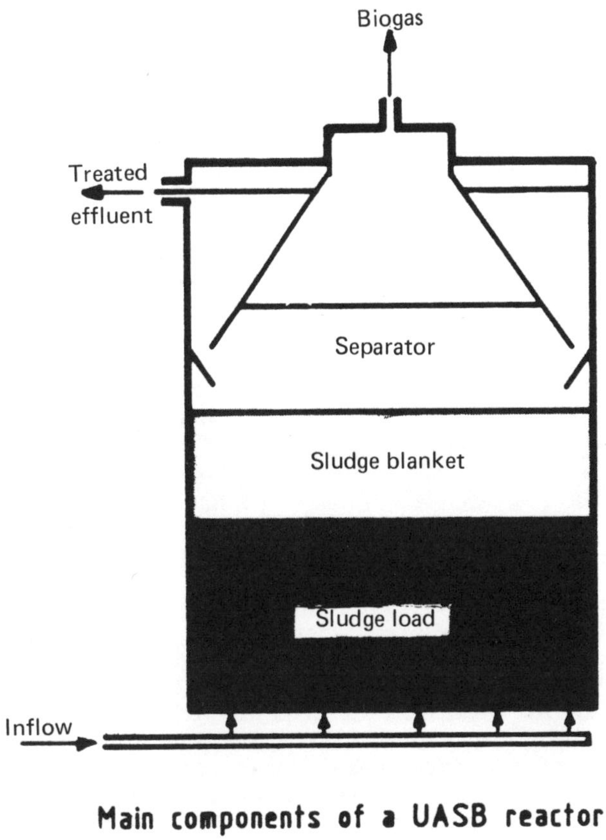

Figure 1. Main components of a UASB reactor

Attainable Loadings and Design Considerations

The attainable loading of any biological system depends on the amount of active sludge in that system and its properties of retaining the sludge under the required process conditions. In this respect the UASB process has proved in practice that, due to the development of sludge with excellent settling characteristics, high sludge concentrations can be attained. In full-scale reactors, sludge concentrations of 10-20 % are common for the sludge bed (approx 1/3 of reactor volume) which, together with specific sludge activities of 0.5-2.0 g COD/(g VSS.day), gives the reactor an enormous potential in respect of its metabolic activity.

Loading rates of 60-70 kg COD/(m^3.day) are reached on an ideal substrate as a VFA mixture. The presence of highly active sludge is one condition for high-rate performance which must be met; another condition is an optimal contact between substrate and sludge.

This brings us to the subject of influent distribution, mixing by biogas production, and the consequences of shut-down periods. In general the design of the influent distribution system is not critical when reactor loadings exceed 6 kg COD/(m^3.day). Measurements indicate that the gas production from these loadings results in a completely mixed reactor, thus enabling optimal sludge/substrate contact. Under these conditions one inlet point per 5-8 m^2 will suffice. During start-up the gas production is limited and the feed distribution system is of more importance for the sludge/substrate contact. To shorten the start-up period a larger number of feed inlet points is therefore desirable. This also applies for low-loaded systems in cases when the biodegradibility of the substrate is limited or treatment is at suboptimal temperatures (e.g. slaughterhouse and municipal wastewater).
In practice such considerations as short shut-down periods are also of importance (e.g. 5-day-week operation). To minimize the required restart period, the sludge must be prevented from settling out completely. This can be effected by a constant recirculation (maintaining the "normal" hydraulic conditions).

The Three-phase Separator

The main condition to be fulfilled in the design of the three-phase separator is the accomplishment of an effective separation of the gas before the water/solids mixture reaches the settling zone of the reactor. The gas collectors installed under the settler walls are responsible for this separation. The surface of the gas/liquid interface created in these gas collectors is a critical design criterion. To prevent the formation of a layer of floating sludge, the dimensions of the three-phase separator should be such that the gas-liquid interface in the gas collector is well mixed by the upflowing gas. On the other hand, the gas-liquid interface should not be too small, because it might lead to severe scumming. General design criteria for the three-phase separator can not be provided, as the problem of unacceptable scumming, for instance, is also related to the wastewater composition and the loading conditions of the reactor. Each separator has to be designed to meet the requirements of each specific project.

The second objective is to entrap the sludge flocs/granules conveyed into the settler compartment with the upflowing water, and return them to the digestion compartment. If flocculant sludge is expected to develop, the maximum surface loading of the reactor is limited to approximately 2 $m^3/(m^2.h.)$, to provide adequate sludge retention. In such a situation the substrate level in the settler compartment, clearly an operational aspect, may also become critical as gas production in that area will hamper sludge retention. In case of granulation, higher surface loadings can be tolerated.

Start-up

With respect to the start-up of UASB reactors it has been found that volumetric loadings exceeding 5 kg COD/(m^3.day) can be accommodated within 8 weeks when digested sewage sludge is applied as seed material. In this period the sludge characteristics will change dramatically, in respect of both the specific activity and the settling characteristics. Table 2 gives guidelines for the start-up of a UASB reactor with digested sewage sludge as seed material.

Table 2. Tentative Guidelines for the Start-up of a UASB-PLANT (Lettinga et al., 1984)

1.	Amount of seed sludge: 10-20 kg VSS/m^3
2.	Initial sludge load: 0.05-0.1 kg COD/(kg VSS.day)
3.	The space load should not be increased unless all VFA are well degraded (over approximately 80 % reduction for medium strength wastes)
4.	Voluminuous sludge should be allowed to wash-out; the heavy part of the seed sludge should be retained

By obeying these guidelines it will be possible to cultivate sludge with good settling characteristics in a period of 2-3 months (Figure 2). In most cases this will be granular sludge.

Besides the physical sludge properties, the specific sludge activity will also improve. In this respect sewage sludge has an activity of about 0.05-0.10 g COD/(g VSS.day), while this will increase by a factor of 7-15 when the sludge is fully adapted.

By using granular sludge from other UASB reactors as seed material, the start-up period can be limited to less than 1 month, depending on the nature of the waste to be treated, the origin of the seed granular sludge and the quantity of seed sludge. When a stable process operation is acchieved the loading rates indicated in Table 3 can be reached in the treatment of various wastes.

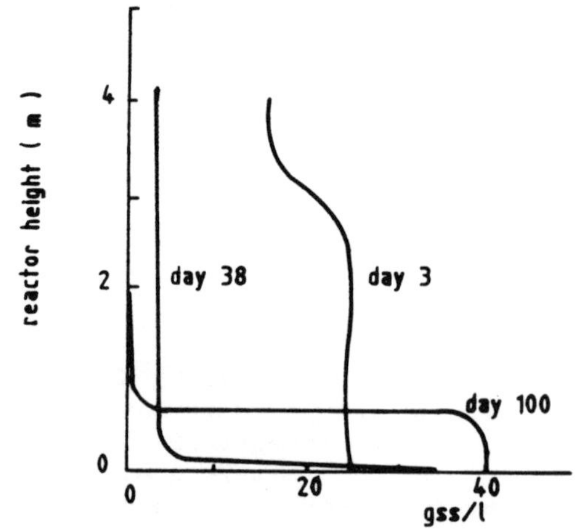

Figure 2. Sludge Profiles over a UASB reactor Height at Day 3, 38 and 100 using Digested Sewage Sludge as Seed Material (Hulshoff Pol et al., 1983)

Table 3. Attainable Loading Rates with the UASB process

Fullscale		
Starch (wheat, maize, potato)	10-30	kg COD/(m^3.day)
Sugar	20-30	"
Slaughterhouse	5-10	"
Dairy	5-20	"
Paper	5-10	"
Laboratory		
VFA	50-60	"

The previously described considerations in reactor design are below illustrated by a number of project descriptions.

PROJECT DESCRIPTIONS

Case: AMPI (USA)

Introduction. The cheese factory at Hillsboro (Wisconsin) is one of the 43 production plants of Associated Milk Producers Inc. (AMPI). It processes milk into mainly New Holstein Münster cheese. The whey, which is a waste product of the cheese production process, is being hauled away for drying at AMPI's expense. Furthermore, the actual discharge loads in relation to the discharge requirements, forced the corporation to investigate the feasibility of anaerobic treatment of its waste flows. Grontmij, in co-operation with Robex (Madison) and Mr. G. Richards (Milwaukee) carried out a pilot investigation programme.

Basic design data. The raw effluent of the factory consists of whey mixed with the wash and cleaning water. Table 4 summarizes the composition of the whey and the total waste flow.

Table 4 Wastewater characteristics - AMPI (USA)

		Whey	Total flow
Flow	m^3/day	200	420
COD	g/l	55	32
BOD	g/l	33	19
TKN	mg/l	1100	580
P-tot	mg/l	650	310
pH		4.5-5.5	5.0-10.6
Protein	%	0.6-0.9	-
Lactose	%	4-5	-
Fat	%	0.04	-

The waste flow's high concentration and the nature of its organic compounds make it an excellent substrate for anaerobic treatment.

Due to the high COD of the wastewater, mixing with treated effluent is favourable for proper reactor operation. As the nutrient concentrations are sufficient to allow anaerobic treatment, no dosing of nutrients or trace elements is required.

Results. The pilot UASB reactor was seeded with highly active granular sludge from Agropur Bon Conseil in Canada. Table 5 shows the performance during the pilot study.
From this table it can be seen that on this substrate high reductions of COD, and consequently high production rates of biogas, were achieved. The UASB reactor was operated with a hydraulic retention time of 16 h (with recycling). These favourable results have led to the design of a full-scale treatment plant, a process flow diagram of which is presented in Figure 3.

Table 5. Performance of the UASB Reactor during Pilot Study

		Influent	Effluent	Reduction
COD	mg/l	33000	5.600	86 %
BOD	mg/l	14500	680	96 %
Biogas production				0.50 m^3/kg COD
Biogas composition				64 % CH_4
Sludge Settling				25-54 m/h

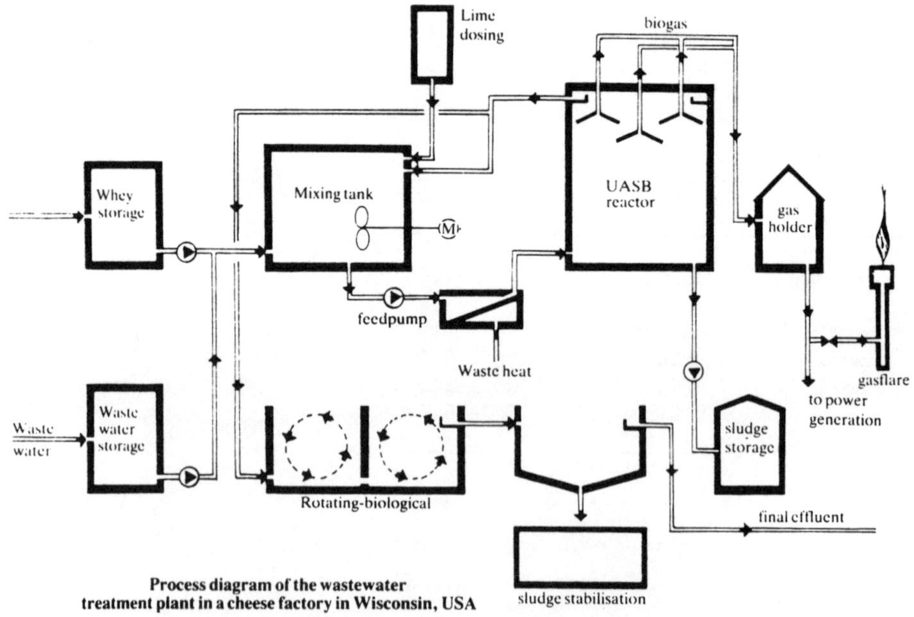

Figure 3. Process Diagram of the Wastewater Treatment Plant for a Cheese Factory in Wisconsin, U.S.A.

The main part of the design will be a UASB reactor. The lime dosing (as indicated in Figure 3) will be used to control potential problems of acidification in the event of a disturbance in reactor performance. In normal reactor operation, no lime dosing is required because of the high buffering capacity of the feed to the UASB (a mixture of influent and recycled effluent). In order to reach the final effluent requirements, aerobic post-treatment (rotating biological contactor) has been included in the design. After aerobic treatment, the BOD of the raw effluent will be reduced from 14,500 mg/l to 19 mg/l, a reduction of over 99 %.

Case: Wheat Industries (Ireland)

Introduction. In 1984, soon after construction of a wheat starch factory in Cork, Ireland, had been completed and the factory had been commissioned, an anaerobic treatment plant was started up to treat the factory's wastes. Several waste flows are generated in the production process. In the design phase of the production plant it became evident that the effluent from the starch extraction plant would not meet the effluent requirements set by the water authority. Analysis of the various discharge flows showed that pre-treatment upon discharge of this separate stream would reduce the total pollution load to the required level. Table 6 indicates the anticipated characteristics of the starch plant effluent.

Table 6. Wastewater Characteristics of a Wheat Starch Factory

Flow	m^3/h	35
COD	g/l	20.2
BOD	g/l	12.6
SS	g/l	1.5-3.0
TS	g/l	10-15
TKN	mg/l	45
P	mg/l	4
Temp.	°C	20
pH		5-6.5

Design criteria. The design of the treatment plant is based on a required reduction of 85 % in COD and 90 % in BOD. As the wastewater treatment plant was designed and constructed before the wheat production plant was built, only approximate data were available on both the quality and the quantity of the raw effluent.
For that reason the starting point was a rather conservative design load of 7.5 kg COD/(m^3.day). Although the origin of suspended solids guarantees relatively good biodegradability, the expected concentration ranges were an important design consideration. Potential shock loads of solids might threaten the stability of the methane reactor.

A build-up of solids in the reactor, a phenomenon more likely to occur during periods of limited gas production (start-up), can lead to a sudden acidification of the reactor content. For that reason a pre-acidification step was incorporated into the plant design. The pre-acidification tank consists of a completely mixed tank without sludge retention. The pH is adjusted by NaOH dosing. The performance with respect to hydrolysis of solids is indicated in Figure 4.

An optimal growth of sludge is a first requirement during the start-up of anaerobic reactors. This optimal growth requires an intense contact between the sludge and the substrate (influent COD). As the internal mixing in UASB reactors depends on the influent distribution and the gas production, the influent distribution is of the utmost importance during start-up, when gas production is minimal. This consideration was especially important as no anaerobic sludge was available in Ireland. Only a minimal amount of 150 m^3

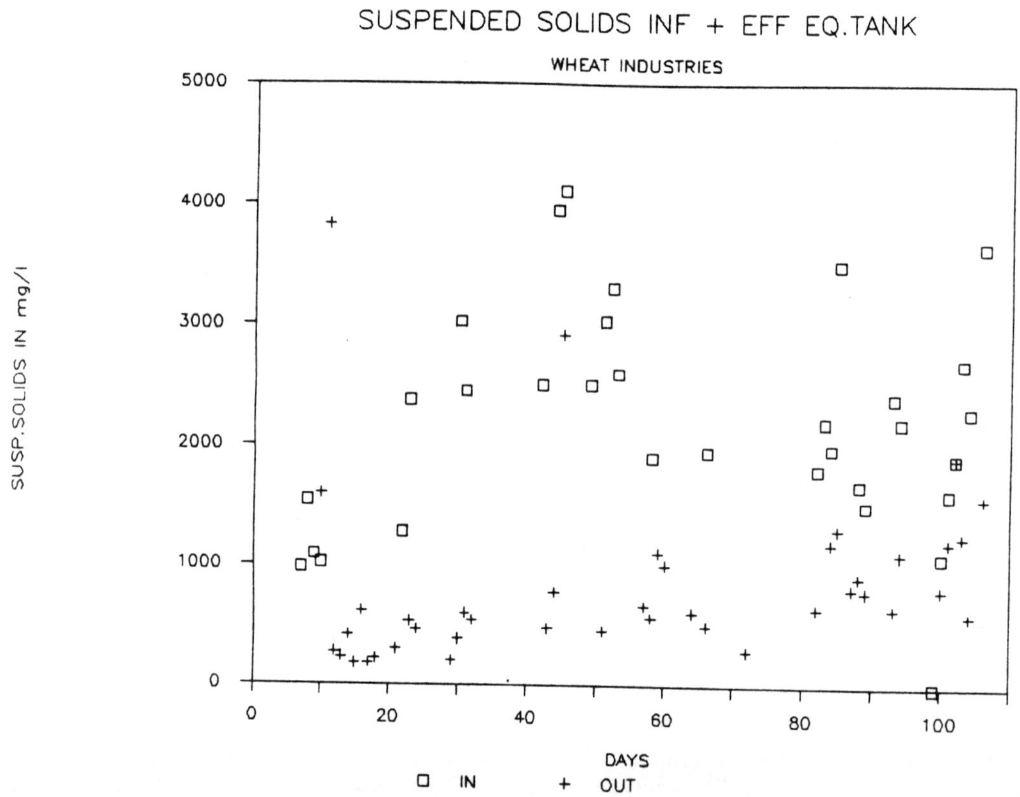

Figure 4. Hydrolysis of suspended solids.

was imported from Holland to start up this 2400 m^3 plant. Figure 5 presents a process flow diagram of the treatment plant.

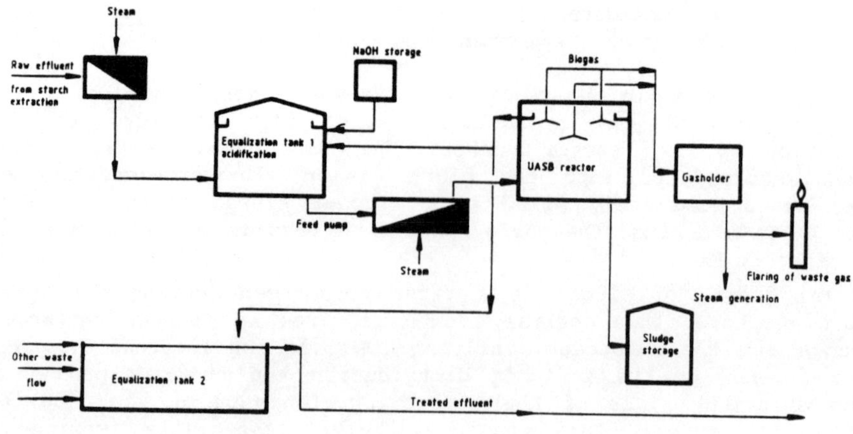

Figure 5. Flow Sheet of the Anaerobic Treatment Plant at a Wheat Starch Factory (Ireland)

The raw wastewater is collected in the equalization acidification tank, where the pH is corrected. After passing through a second heat exchanger, the wastewater is pumped into the UASB reactor. The treated effluent flows into a second equalization tank in which it is mixed with the other waste flows of the factory upon final discharge. The treatment plant includes a sludge buffer tank with a volume large enough to allow for restart within a few days upon a possible calamity.

Start-up and operating results. To startup the installation, the reactor was filled with active granular sludge from other treatment plants. Thanks to this sludge, a rapid and steady increase in the loading of the reactor has been achieved, which is illustrated by the increase in methane production shown in Figure 6.

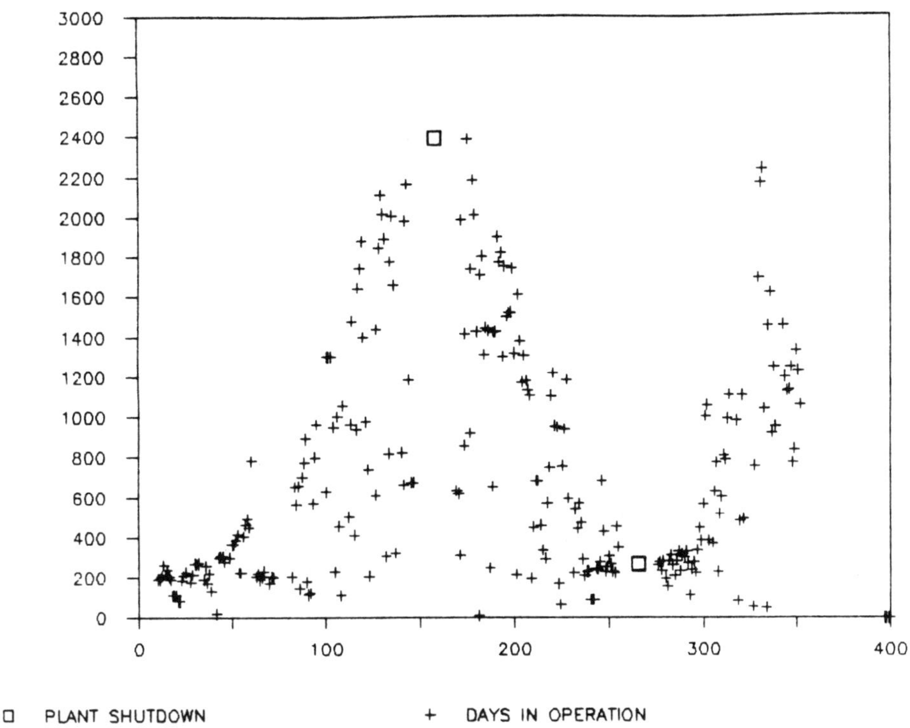

Figure 6. Methane Production upon Start-up of the UASB reactor at Wheat Industries, Ireland

The COD reductions varied in this stage from 70 to 85 %. Even after a shut-down period of several weeks the treatment plant could be loaded at design loads almost immediately upon restart (shut-down a round day 150 in Figure 6).

However, starting from day 200 the COD-reductions and methane production decreased and subsequently the loading rates of the reactor had to be lowered to unacceptable loadings of 1-2 kg COD/(m^3day). An investigation program was initiated and both in Ireland and the Netherlands detailed analyses were carried out on sludge properties, wastewater characteristics and treatment plant performance. From this research it was concluded that the specific sludge activity had fallen dramatically during a period of 100 days. Normal sludge activities as experienced initially were approximately 0.8–1.0 g COD/(g VSS.day) with the influent as substrate. This activity finally decreased to a level of 0.1 g COD/(g VSS.day).

As no inhibition of the sludge by toxic components could be determined, it was finally shown that a trace element deficiency was responsible for this phenomenon. The required trace elements and concentrations were determined, and a solution of them was dosed to the influent.
With these measures the treatment plant regained its design efficiencies and methane production increased again to normal rates in no more than two months (Figure 6). It should be noted that this was done without the addition of new granular sludge to the reactor. At the time of writing the COD reduction varies from 80 to 90 %.

Case: ZBB Koog aan de Zaan (The Netherlands)

Introduction. The ZBB maize starch factory produces starch, glucose and other starch derivatives from maize. The production process leads to highly concentrated wastewater which has until recently been discharged into the Zaan river. As a result of the high discharge rates the possibility of pre-treatment was evaluated. As only limited space was available, anaerobic treatment was a logical solution. The treatment results will indicate that the choice of this technology was also correct with respect to the overall performance.

Basic design data. Table 7 summarizes the main wastewater flows and the characteristics of the total waste flow.

Table 7. Wastewater characteristics maize starch factory.

		starch	glucose	derivatives	total
flow	m^3/h	40	5	4	40–50
COD	g/l	9	1,5	30	5–26
TKN	mg/l	–	–	–	50
Ptot	mg/l	–	–	–	2
pH	–	–	–	–	5–6

As the production process is partly batch-wise, large fluctuations in salt, COD and sulphite concentrations, may occur. To ensure a proper operation of the process, a large equalization volume is necessary. The wastewater can also contain peak loads of undissolved starch particles that can lead to a decrease of process stability of the anaerobic reactor. By installing a buffer tank in which the pH is controlled by addition of lime, an equalization of these peak loads can be established, while the starch particles will be partly hydrolized and converted to fatty acids. The temperature of the wastewater does not require any adjustment. Because of the relatively low content of nutrients, dosing of nitrogen and phosphorus is required. Figure 7 presents a process flow diagram of the treatment plant.

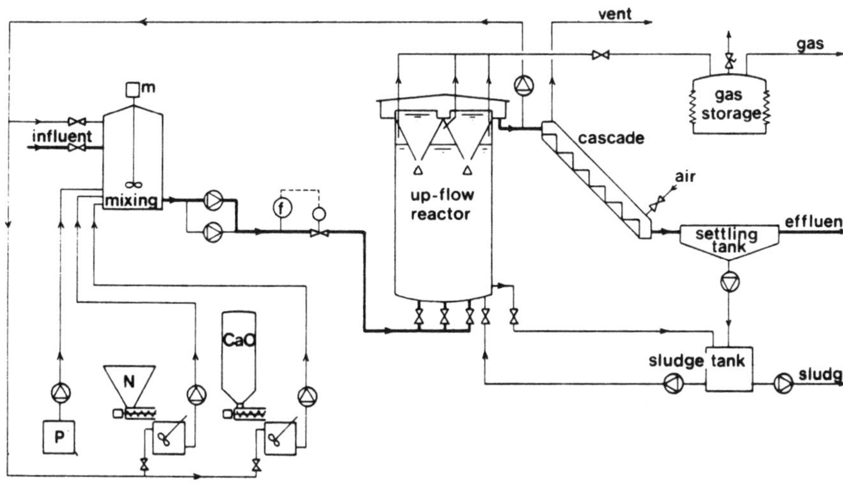

Figure 7. Process Flow Diagram of the Anaerobic Treatment Plant in the ZBB Maize Starch Factory

The wastewater flows are collected in the buffer tank, which has a volume of 1000 m^3. In this buffer tank the pH is controlled while nitrogen and phosphorus are added to the wastewater. The wastewater is pumped into the anaerobic reactor and the final effluent passes a cascade. Finally the effluent passes by a settler before being discharged into the city sewerage. This settler was installed to control the possible wash-out of sludge particles, but it is not used for this purpose, as losses of suspended solids from the UASB-reactor are minimal. The biogas is collected in a gasholder and is fed into the boilers of the factory without treatment.

Performance data. The performance of the plant over a period of about 4 months is shown in Figure 8. Evidently the reactor's design capacity of about 11 kg COD/(m^3.day) is frequently exceeded. Nevertheless, the efficiency of COD removal by this UASB reactor is normally in the 90-95 % range. Even high peaks in the daily load of the reactor, up to values of 30 kg COD/(m^3.day), have little effect on efficiency. These figures illustrate the stability and applicability of the UASB system in the treatment of wastewaters from starch production.

The excellent results are due to the high sludge concentrations that can be maintained in the reactor. Measured values of organic dry solids range from 2 % just below the three-phase separator to 5 % near the bottom of the reactor. The amount of biogas produced is more than 800,000 m^3/year. The gas contains approximately 70 % methane.

The treatment plant is shut down during weekends. When the reactor is started up again, the plant is at full capacity in a few hours and the gas production rate reaches its maximum (see Figure 9).

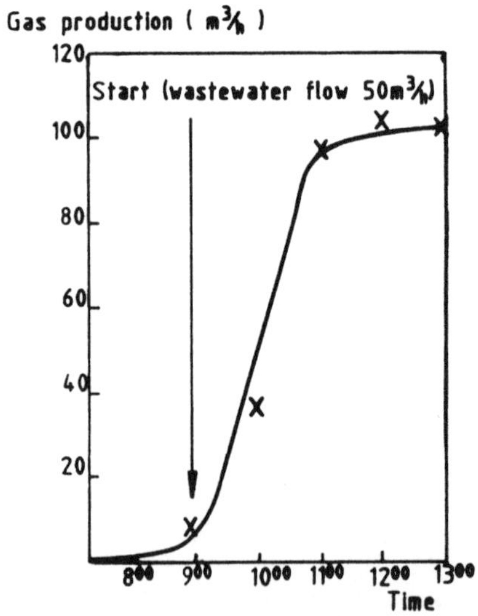

Figure 9. Gas Production Rates during Weekly Restarts

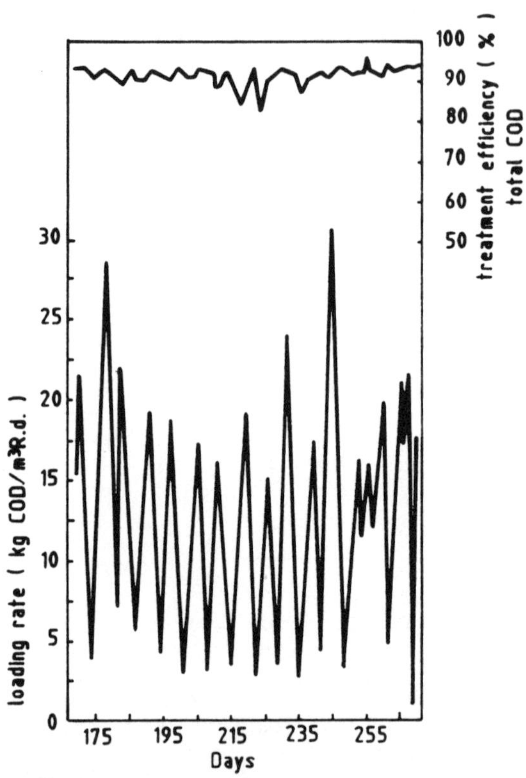

Figure 8. Treatment Efficiency and Loading Rate

Case: Leachate Treatment (The Netherlands)

Introduction. Leachate from sanitary landfills is often heavily polluted. In 1980, the Dutch government ministry concerned with environmental affairs commissioned Grontmij to investigate the anaerobic treatment of such waste in a pilot plant.

The quality of the leachate varies depending on the age of the waste through which it percolates. In relatively young landfill sites the leachate can be highly polluted with low-molecular-weight organic acids and by ammonia, chloride, sulphate and heavy metals. After a period of 2-4 years the metabolic activity of the landfill itself will increase considerably, resulting in a drop of COD of the leachate from 25,000 mg/l to 2,000-5,000 mg/l. The COD/BOD ratio increases and the organic pollutants now mainly consist of long-chain carbohydrates which are difficult to degrade. For the treatment of this wastewater, nonbiological processes such as reverse osmosis are suitable. In the pilot investigation the anaerobic treatment of leachate of a young, not yet methanogenic, controlled tipsite was investigated.

Table 8 summarizes the results of this test. The reactor was operated at an average loading rate of 7.3 kg COD/(m^3.day) and at a temperature of 27 °C. The biogas produced contained 85 % methane.

Table 8. Anaerobic Treatment of Leachate in a Pilot Plant

	Raw wastewater (mg/l)	Final effluent (mg/l)	Reduction (%)
COD	4600	720 *	81
TKN	190	175	13
Fatty acids	1930	380	77
P-total	6.8	1.6	67
Cr	0.13	0.06	63
Ni	0.12	0.05	59
Cu	0.08	0.06	54
Zn	1.14	0.19	80
Pb	0.02	0.006	67
Cd	0.004	0.003	67
Fe	155	36	69

* COD/BOD ratio of final effluent 6-9

The experience gained from these studies was used in the design of a leachate treatment plant at Bavel.

Design data. At a landfill site at Bavel, The Netherlands, the leachate is collected in a drainage system above a water-impermeable foil. The leachate is recirculated over the solid waste. Surplus leachate has to be dicharged to surface water. In order to meet the discharge requirements (Table 9) a two-stage wastewater treatment system was designed incorporating a UASB pre-treatment and a post treatment according to the reverse osmosis process.

Table 9. Wastewater Characteristics and Discharge Requirements at Bavel

	Leachate mg/l	Discharge requirements mg/l	Required Reduction rates %
COD-total	26000	75	99.7
TKN	1330	4	99.6
BOD		5	
NH_4-N	1300	3	99.8
Fatty acids COD	18000		
pH	6.6-6.8		
SS	100		

As can be seen from the composition of the wastewater, it originates from a young section of the tip site. Because of the high concentrations short-chain fatty acids, direct treatment by reverse osmosis would not result in the required reduction rates. The leachate will therefore be pre-treated in a UASB reactor. It is known that with time every section of a landfill will become methanogenic. The anaerobic stabilization of the fatty acids will then take place in the landfill itself. The resulting "methanogenic leachate" can be directly treated by reverse osmosis. However, leachate from this section of the Bavel landfill requires anaerobic pre-treatment for several years.

Figure 10 shows the set-up of the designed two-stage treatment plant.

Results. Three month's after start-up, the 60 m^3 UASB reactor and the reverse osmosis unit treated 2 m^3 of leachate per hour. The UASB reactor operates at a loading rate of 15 kg COD/(m^3.day). The 80-90 % breakdown in COD resulted in 400 m^3 of biogas/day (72 % CH_4).

The reduction in heavy metals amount to over 99.8 % for the two-stage set-up. The final effluent is discharged to surface water.

Table 10. Treatment Results at Bavel

	Leachate mg/l	After UASB mg/l	After reverse osmosis mg/l
COD	26000	4000	6
TKN	1330	1270	15
NH_4^+-N	1300	1260	13
Fatty acids COD	19000	2300	
SO_4^{2-}	380	97	
Cd	0.02	0.004	< 0.1 ppm
Cu	0.03	0.02	< 0.1 ppm
Hg	0.02	0.002	< 0.1 ppm
Pb	0.11	0.06	< 0.1 ppm

Figure 10. Flow Diagram of Leachate Treatment at Bavel, The Netherlands

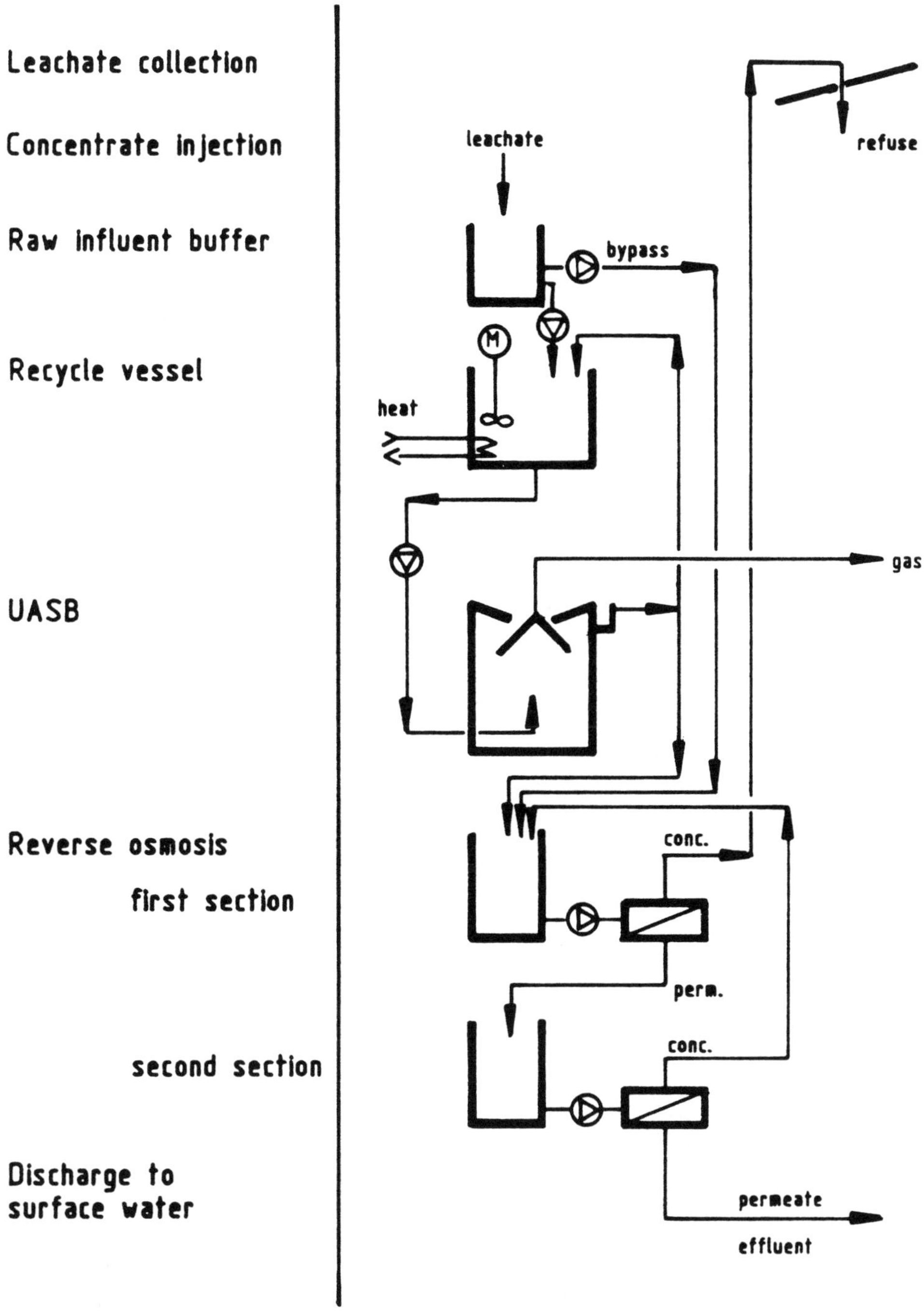

Case: Venco Naarden (The Netherlands)

Introduction. Venco is a liquorice factory that produces 6000 t of liquorice a year. The wastewater of the production process is discharged into the city sewer. In 1983 the discharge taxes amounted to over US$ 40,000. Due to these high taxes Venco commissioned Grontmij to study the potential of wastewater treatment. Anaerobic treatment according to the UASB concept appeared to be the most feasible of several possible alternatives.

In this specific case the scale of the project did not allow for extensive pilot testing, so the design of the full-scale anaerobic treatment plant had to be based on limited laboratory research and our experience originating from the anaerobic treatment of other waste flows, for instance in the starch industry (Jans et al, 1986).

Basic design data. The liquorice is produced using a syrup. Tabel 11 gives the overall composition of this syrup and of the wastewater. The pollution load consists mainly of carbohydrates and protein and, to a lesser extent, fat. The protein contributes 15-30 % of the total COD.

Table 11. Wastewater Characteristics of Liquorice

		Liquorice syrup	Total flow
Flow	m^3/day	-	50
COD	mg/l	-	12000
BOD	mg/l	-	5800
TKN	mg/l	-	260
P-tot	mg/l	-	-
pH	-	-	-
Protein	%	3.9	-
Starch	%	31	-
Sugars	%	31	-

The wastewater is easily biodegradable and contains no toxic substances. The relatively high amount of proteins required special attention in order to increase process stability. For instance, to prevent coagulation of the proteins in the system the residence time of the wastewater in the sewerage system was reduced from an average of three hours to approximately 30 minutes. Furthermore it is known that proteins may cause problems in the UASB reactor because of sludge flotation and build up of scum layers. The three-phase seperator was designed to counter these problems.

The wastewater is produced in an eight-hour day, five days a week, but since continuous operation increases the stability of the anaerobic process and reduces the size of the required reactor, a buffer tank of 120 m^3 was incorporated in the design. Additional advantages of this buffer tank are:

- reduction of final discharge taxes because discharge is continuous;
- stable reactor performance due to buffering of large fluctuations in the COD-load brought about, for instance, by cleaning activities.

Laboratory testing showed that in 23 hours, 32 % of the COD could be converted into mainly acetic acid and propionic acid by acidifying bacteria. Caustic dosing was needed to control the pH in this buffer tank. By recirculating the anaerobic effluent over the buffer tank, a considerable reduction of this dosing can be achieved. The anaerobic reactor has a volume of 50 m^3 and is designed according to the Upflow Anaerobic Sludge Blanket process at an average loadin grate of 10 kg/(m^3.day). The treatment plant also includes a sludge buffer with enough anaerobic sludge to allow for a restart within a few days of a calamity. Figure 11 gives a process flow diagram of the treatment plant.

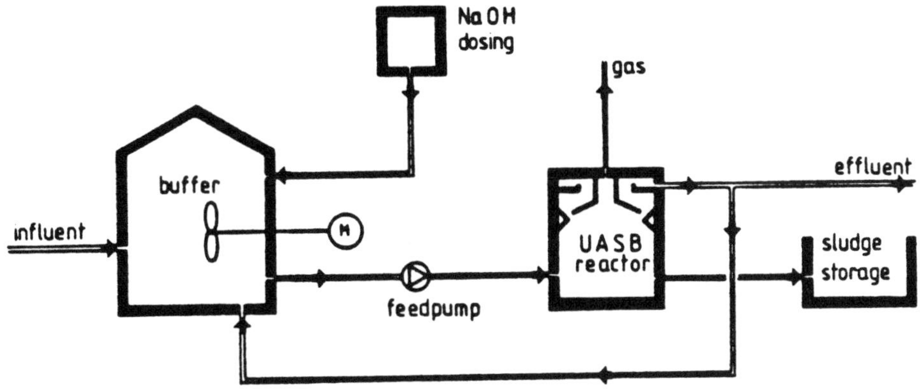

Figure 11. Process Flow Diagram of Treatment Plant in the Venco Liquorice Factory

<u>Start up and first operating results</u>. The full-scale treatment was completed by January 1985. Shortly after start-up with granular seed sludge, the treatment plant could be loaded according to the design data. Figure 12 presents the COD reduction for March and April 1985, calculated on total COD in the raw effluent and in the final effluent.

After 6 months of operation there was a decrease in specific sludge activity. An investigation showed that there was a stortage of certain trace elements. Dosing of these elements resulted in the resumption of normal process operations.

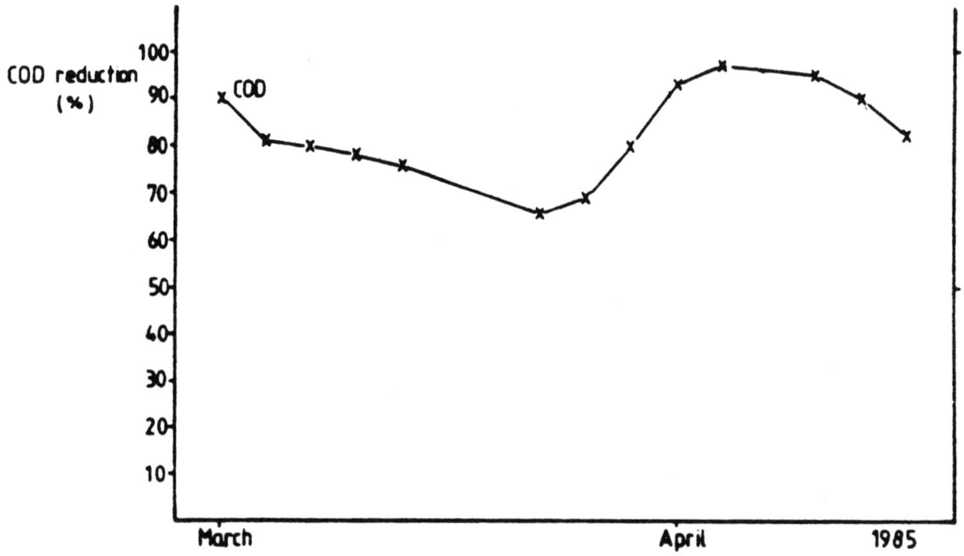

Figure 12. Performance of Wastewater Treatment at Liquorice Plant (Venco, NL)

SUMMARY

Anaerobic treatment of wastewater may now be considered a "grown-up" technology, as can be seen from the results of many full-scale operating reactors. Especially the UASB process, developed in The Netherlands, has been successfully applied in the treatment of a large variety of wastwaters. It has been possible to achieve a large degree of process stability at minimum investment and operational costs as was illustrated by a number of projects descriptions.

REFERENCES

Jans, A.J.M. and Piscaer, P. (1986).
Anaerobic treatment of wheat starch effluent. Presented at conference: Anaerobic Treatment a Grown Up Technology, 15-19 September, Amsterdam.

Hulshoff Pol, L.W., Zeeuw, W.J. de, Dolfing, J. and Lettinga, G. (1983).
Start-up and sludge granulation in UASB reactors. In: Proceedings of the European Symposium AWWT, 23-25 November, Noordwijkerhout (W.J. van den Brink, Ed.), pp. 40-43. TNO Corporate Communication Department, The Hague, The Netherlands.

Lettinga, G., L.W. Hulshoff Pol, I.W. Koster, W.M.Wiegant, W.J. de Zeeuw, A. Rinzema, P.C. Grin, R.E. Roersma and S.W. Hobma.
High-rate Anaerobic waste-water treatment using the UASB Reactor under a wide range of temperature conditions.
Biotechnology and Genetic Engineering Reviews, (1984), 253-284.

ANAEROBIC MARKETING–THREE CASES IN POINT

E.A. Richards

Bioenergy Systems, Ltd.
Milwaukee, Wisconsin

ABSTRACT

Described briefly are three BSL/Robex/Grontmij UASB (Upflow Anaerobic Sludge Blanket) dairy industry marketing projects of the recent past, along with some problems encountered during the marketing phase. These proposed projects, in order of submission, included the following:

1. A UASB system only, releasing effluent to a large municipal plant.

2. A UASB system with a low-level aerobic post-treatment for release of effluent to a small-town municipal plant.

3. A UASB system with high-level aerobic post-treatment to bring BOD and TSS to under 20 mgl for direct release to the environment.

Along with the present status of the above three projects, some of the specific problems arising from the liability, pricing, bacteriological, and legal aspects of these types of projects are discussed, with some limited recommendation for problem approach.

Thus far the conference has been proceeding on a generally lofty intellectual, professional, and technological plane, but that high level approach may abruptly cease right here inasmuch as we will be discussing the vagaries of sales and marketing.

In going through the attendance list last evening I noted that there were about fifteen potential anaerobic system buyers and the rest of us were sellers, and I thought God help those fifteen buyers because at last we have them where we want them - although viewing the gathering here this morning it seems as though some may have escaped overnight.

My brief discussion is intended to illustrate some of the problems in the dynamic marketing of anaerobic systems, and while preparing the text over the weekend it came to me that those marketing problems with regard to anaerobic systems were not too much different from marketing problems encountered in selling any sort of system involving design and build process projects. They are all difficult in the main - tough, tough selling jobs.

But like the kids say, life is tough and then you die!

So let us get to the primary topic; and please note that this paper is not an attempt to teach marketing techniques, but is only a tongue-in-cheek appraisal of small firm's efforts to sell anaerobic systems to clients in the dairy industry.

A time arrives in an engineer's life when he must consider more than just the design package of nuts, bolts, wires, etc., in a given field, and become aware of the process involved in taking that package and selling it to a customer who will in turn pay him for it. The process usually does not take place without a great deal of effort on the part of someone who must convince a client that he really needs this particular item, and that its use will bring the client fame, glory, prestige - and a hopefully a pile of money.

The group that works this wondrous process may have the loose title of "Marketing Division," and is made up of the Marketing managers and the outside Sales Representatives. In order to be successful, a triad made up of Engineer, Marketeer, and Sales Rep should be a coordinated amalgam, a mix which takes the talents of each and forms a mighty force to design, build, sell, and install a given project.

Anyway, that's how it should be....

So for purposes of clarification, let us use the old saw to define the roles of those named above:

A Sales Rep is a man who knows a great deal about very little, and who goes along learning more and more about less and less until finally he know practically everything about nothing.

A Marketing Manager, on the other hand, is a man who knows very little about many things, and keeps learning less and less about more and more, until he knows practically nothing about everything.

An Engineer starts out knowing everything about everything, but ends up knowing nothing about anything, due to his association with Marketing Managers and Sales Reps.

So it's apparent why they can work well together.

But even the best of such corporate teams, with a good product to offer, still can have problems in marketing its wares. And it is some of these marketing problems in the field of anaerobic treatment that I am here to discuss this morning. And the way I propose to do this is by relating the sequences of events transpiring in three particular anaerobic treatment marketing endeavors, the outcome of those efforts, and then describing some of the problems encountered.

In 1983, my firm of Bioenergy Systems set up a working relationship with Robex Ltd of Madison to market anaerobic treatment systems for the dairy industry. Dairy effluent, with its carbohydrate-rich character, is an ideal substrate for anaerobic treatment. Cheese whey composes the bulk of the dairy effluent, and in its most treatable form, after deproteinization when it becomes whey permeate, has a COD content of about 60,000 mgl, and offers the capability of generating significant amounts of methane gas useable for energy as boiler fuel or running electrical power generators.

The production of such energy from byproduct should be a marketing option attractive to any corporate food processor.

CASE I:

We had made the decision that our anaerobic thrust would be based on the tried and proven Upflow Anaerobic Sludge Blanket technique, the UASB system.

Working in conjunction with Ton Jans, Msc., an engineering colleague from the consulting firm of Grontmij nv, in the Netherlands, we installed a modest pilot plant operation in a central Wisconsin dairy. The arrangement was that we would operate the pilot plant in return for our basic expenses in the hope that if successful, the dairy would purchase an anaerobic facility of our design and construction.

What innocents we were!

In this dairy, the effluent consisted dairy washwater and whole whey. In brief, the pilot plant was run batch mode using a source tank provided by the dairy. Our Robex colleague, Mr. Rajan Vembu, was at the plant day and night to oversee the operation, running the plant and making all the ph, COD, and methane tests. Because the pilot work was continuous, the dairy also had one of their night crew look in on the process at intervals, if only to push a button or turn a valve handle, just to see what happens.

Results of the pilot study indicated anaerobic reduction effiencies approaching 92%; methane was generated at the rate of about 5 cubic feet per pound of COD converted. It appeared to be a very successful and conclusive pilot program.

On the basis of those program results, we prepared a comprehensive proposal, with site drawings that contained all of the test data, costs, scheduling, equipment, vendors chosen, income projections, with the final price to the dairy of just over a million dollars, and a payback (including all R&D, Tax, and Energy Credits) of 2.2 years, a figure deemed within acceptable limits by the bean counters of the world.

(Incidentally, the site chosen by the dairy for our facility was a small, pie-shaped area in which we did manage to squeeze all of the building, equipment and tanks)

The proposal was submitted and the company response was that " they would get back to us."

After a decent interval of nail-baiting on our part, and no response from the potential client, we followed up a number of times with the company and were told the document was still under study by their analysts.

Time passed slowly, as it usually does when awaiting an answer, with frequent calls by Mr. Vembu, who was told the documents were still 'under study.' Subsequent calls to the company were returned with what seemed a growing reluctance on their part. We were always told that a decision had yet to be made, but was imminent.

During a trip to the city in question, we had occasion to drive past our proposed job site and were mildly surprised to see that a large milk tank was being built right where the main anaerobic reactor would have stood.

This we took as a somewhat negative response on the part of the dairy.

When management was at last pinned down to a go/no-go decision, they said that while the results of the tests showed the anaerobic project to be of more than adequate viability, they were thinking of moving the plant to another location and until that decision was made no anaerobic plant would be built until after a move.

At last word, the dairy has purchased the rest of the block on which they were located, and has constructed another warehouse.

CASE II.

Bloodied but unbowed, we undertook another pilot plant operation at a large dairy in west central Wisconsin. On this occasion we provided a pilot plant operation somewhat more sophisticated than the one previous, and Grontmij nv provided a skilled plant operator, Mr. Hans Draayer, who spent many hours at the site running the process, making the necessary tests, and recording all of the pertinent data.

In this case, however, we were astute enough to charge a modest sum for our services in the hope we would break even on the pilot project. This was a hope not to be realized.

The dairy, and its local manager and staff, gave us their complete cooperation and assistance in all respects, such as the procurement of effluent for the various phases of the pilot project, even to the point of providing personnel to aid in set-up and operation.

Results of the pilot study were again extremely positive.

Reductions of the effluent BOD/COD reached 95% on occasion, averaging 90% most of the time. Methane was generated at the rate of 5.2 cubic feet per pound of convertable COD, a total biogas of 8.13 cubic feet per pound, of which methane was about 64% and carbon dioxide about 36%. H2S in the gas was negligible.

It seemed that we had everything going for us in this project. First of all, the pilot plant itself was housed in a building formerly used as an airplane hanger, and was located near the end of a runway. We were able to fly our small rentaplane to the site, land on the runway, taxi over to the hanger, stroll through the dairy laboratory, and arrive at the pilot plant area. It was probably the most convenient remote pilot plant access in engineering history.

Second, adjacent to the dairy was a brand new municipal wastewater treatment plant which was to accommodate most of the dairy loading, and which already had served notice on the dairy that certain limits of BOD level to the muni plant would have to be met if the dairy were to continue high production or contemplate any sort of expansion.

And if our plant gave rise to any tiny emanation of odor it would be lost in the aromas from the municipal plant.

Again, we prepared a comprehensive proposal, this time including provision for a secondary treatment utilizing the activated sludge process, and guaranteeing a maximum BOD to the muni plant. Also included was the option for the dairy to either use the methane for their fuel consumption and sell the surplus, or to co-generate with the utility, selling the surplus KW's.

Our recommendation was to use the energy for co-generation of power, and in that mode, along with the various energy, R&D, and tax credits involved, the anaerobic/aerobic facility had a payback period of 2.7 years.

In this case the dairy in question was owned by a large co-operative, and when we were informed that corporate analysis and study would have to be accomplished before any decision were made, we were prepared.

In the period after the submission of the proposal, several meetings with the parent company were held. These involved informative sessions with high level people to explain in more detail the facets of the proposal, the same type of meetings with middle management to explain the same things, and meetings with the inevitable analysts who, in many cases, could not understand what they were analyzing.

It was at this point that the lawyers entered the scene again. Their original, and blessedly brief, appearance was in regard to the pilot plant contract, and finding that they could not muster enough legitimate objection to that first phase, very grudgingly issued a reluctant OK to proceed.

However, at this juncture, they found more bones to chew.

We had been asked by the company to prepare a contract covering a proposed agreement, which we did, at considerable expense, I might add. Next came the preliminary meeting with their attorneys, who were prepared to battle any sort of reasonable suggestions that the proposed agreement might somehow be fair to both sides.

In spite of this remarkable, but predictable, attitude, the language of the agreement was worked out to the satisfaction of the client and ourselves, if not completely to the satisfaction of the attorneys.

All that was left was the signing of the contract and the work to begin. All was in place: the equipment vendors, the external engineering, the construction contractors, everything.

Time passed....

This time the holdup on the project featured managerial disagreement on final approval, and who would be relieved of making it, along with territorial politics, and why this particular area should get the project and not someone else's, especially since the area in question had once made a minor error in the choice, and ultimate failure, of a $2 million dollar project of another sort.

This merry give and take went on for an extended period without any positive project decision, right up to the day the company announced it was merging with another Co-op, and that all projects were on "hold." But we were given assurance that even if the company did not build the anaerobic-aerobic plant at the proposed site, we would be involved in the design and construction of such a plant at another site.

The company sounded very sincere in this; however, we're still waiting.

CASE III.

This project is on-going, and consists of a Dairy Whey Ultrafiltration plant, along with an Anaerobic-Aerobic complex to produce methane for co-generation, and in the process, reduce the BOD to the point where the cleansed effluent may be issued directly to the environment. It is mentioned only so as to provide a basis for stressing some other factors besides those of marketing and engineering that should be recognized.

And so you will not be held in suspense by those cryptic words, suffice it to say that you may consider our Anaerobic-Aerobic Energy Generation facility as a large, rectangular concrete box, where dairy effluent and whey permeate enter at one point, electrical power is fed back to the power company at another point, and a stream containing less than 20 mgl BOD is emitted to the environment.

The Chicago River should be that clean!

So let's get to some problems in the marketing of Anaerobic systems.

1. THE BUG FACTOR.

Our experience has been that one of the problems has been the unwillingness of the client to accept the fact that this system uses actual living creatures to do the job. If it were simply a chemical process with the attendant changes in chemical structure, there would not seem to be any difficulty. But as long as there are "little bugs" doing the work in ingestive, excretive, and perhaps even fornicatory fashion, the process may be questionable in their minds.

Maybe lurking in the background of their psych is the thought that some day an anaerobic monster will rise out of the sludge blanket and devour everything in sight, including their profit margins.

Other clients, of more prosaic nature, bring up the concern that if the bacteria are indeed living, what is to stop their being killed off by some poison? The fact that good housekeeping in the plant, by their people, would raise the odds considerably against this event happening.

Unstated might be the more realistic fears stemming from realization of the inadequacies of their own labor force. The knob-twiddlers are always with us, the worst being the sweepers and the plant managers.

In any case, the question of bacteria use should be explained in terms of a simple bio-chemical reaction, utilizing a viable element rather than one non-viable. In those euphemistic terms the concept is easier to accept.

2. THE COST FACTOR.

It's difficult for a client to understand, on occasion, the fact that there are certain costs involved in the design and construction of an anaerobic plant that are relatively fixed within effluent discharge ranges. And one of the items most difficult to understand is how the anaerobic plant to be built can approach or cost more than the value of his own plant.

This is a legitimate concern on the client's part.

It can be answered in the main with the production of a breakdown which addresses the various costs for engineering, fees, equipment, construction, insurance and bonding, profit, etc.

Many companies in the anaerobic field prefer to establish a total sum price, with an area to negotiate, rather than to expose their cost breakdowns; however, doing this leaves a number of doubts in the client, who is always concerned that he is being charged more than necessary.

But there are a number of potential anaerobic plant clients who sincerely want and need a plant, for expansion purposes or to meet regulatory agency standards, and who have a solid, well-operated business, but simply cannot obtain the necessary financing to accomplish the project.

This problem can be met in a number of ways, two of which are as follows:

A. Provide a structure whereby the client has the ability to lease without having to expend significant monies. There are numerous lease arrangements which can be utilized.

B. Explore the possibility of having an investor, or investor group, construct a plant on the client property which will not be owned by the client, but will accept all of the client's plant effluent for conversion and subsequent profit by the investor. The investor would guarantee that a pre-set effluent production limit would always be met by the anaerobic plant.

Keep in mind, there is more to anaerobic plant sales than a mere transfer of funds for technology. With a knowledge of the client's business status, and the ability to tailor financing to the client's needs, the potential for anaerobic sales rises dramatically.

3. INSURANCE COVERAGES.

In today's insurance climate, which some have termed an artificial climate of despair contrived by the companies themselves, the fact remains that there must be adequate coverage for the anaerobic or anaerobic/aerobic project.

These can be as follows:

A. Professional Liability and Errors and Ommissions coverage to the value of total project for each of the the engineering companies participating.

B. An Umbrella coverage, including personal injury, Performance and Payement Bonds, etc., to the value of the project for all contractors and sub-contractors involved in the construction process.

C. Insurance covering warrantees for all of the vendors who furnish equipment or services on the project, to the value of the project.

I have used the term "value of the project" for each of the broad categories above and have emphasized it for obvious reasons. If a critical equipment unit has a value of $1500, but a failure on its part would place the entire system in jeopardy, then the coverage must be to project value in order to protect both the client and the anaerobic plant seller.

4. LEGAL RAMIFICATIONS.

The United States today is viewed as a litigous society, especially by many of our friends from Europe, a view accepted by most of us in marketing and sales, who must hustle to keep ahead of the barristerial wolfpacks.

In the second part of Shakespeare's Henry the VI, there is plotting against Henry, and in Scene II, Dick the Butcher says to the would-be king, John Cade, "The first thing we do, let's kill all the lawyers." And Cade replies, "....that I meant to do."

Now I don't subscribe to that philosophy, at least not in total, but I still recall vividly when an ashen-faced witness stumbled out of the court room and told his boss he had just insulted a lawyer. The boss replied, "Good God, how?"

So it should be kept in mind that there is substance to the old charge that attorneys are, many times, deal-breakers, filling the client's mind with so many dire warnings of potential disaster that the client feels the ends don't justify the threats.

But when all the objections of the lawyers are overcome, and it does appear that an agreement will be made, then it behooves the anaerobic system seller to make certain that all of the language in the agreement is specific, concise, and descriptive of the terms.

Each of the clauses in the agreement must say exactly what the client and the seller want it to say, without room for conjecture as to meaning or subject to conflicting interpretations. If the effluent is to have a certain level of BOD after treatment, and if the efficiency of the system is be at a guaranteed level, then the language should so state, with no opportunity for any future misunderstanding.

And when the final agreement language is realized, and the time has come when the parties will become signatory, make certain that the language of the final agreement corresponds with that previously agreed.

Another pitfall. Some client companies have a policy of preliminary agreement to language, but when the final signing is to occur, request further changes in the document beneficial to them, with the implication that without the change no signing will take place.

Case in point: Some time ago an anaerobic plant company made a prelimary agreement with a client to furnish a plant with a guaranteed efficiency of 85%. When it came time to sign the contract, the client had upped the efficiency, slightly, to 95%, with that no-sale threat in the background. The seller, who had spent a great deal of time and money on the project, took the gamble that he could achieve 95% efficiency with his system on a regular basis.

He lost, thus giving rise to the neologistic phrase of "taking a bath in one's own reactor."

Summary.

I have briefly discussed some problem areas associated with the design, construction, and sale of anaerobic systems, and in my summary will offer no clear-cut solution to those problems but only some comments:

A. The anaerobic process always works, no matter the geometry or configuration of the system. The differences are those of method, time, and efficiency. We happen to think the UASB is on of the better systems for many applications.

B. Whether the main point stressed is the ability of the anaerobics system to generate methane, or the capability of the system to minimize municipal loading charges; the marketeer should be extremely conscientious in his approach, and strive for accuracy in his estimates regarding the potential returns for those features. Never misrepresent - be honest to a fault; however, this doesn't mean you should sell yourself short either - if you're good, you're good.

C. Protect your client's interest and your interests by making certain the project agreement spells out terms in exact fashion, and states precisely the conditions of sale. It is good business to do so, and keeps us out on the reactors instead of in the courtrooms.

D. Be adequately insured. It is another protection for both seller and client. Make sure the insurance coverage holds firm at all project levels - design, construction, engineering, contractor, and sub-contractor.

And finally:

Yesterday morning Dr. Speece had referred to the question of what the future holds for anaerobic treatment. In my opinion, anaerobic treatment of waste, no matter the method, no matter the aim - energy or reduction - will be commonplace in the next decade in most industries where a treatable substrate is generated.

Why do I take this position?

Because the bottom line to every industrial venture is ultimate return; and by allowing dollars in the form of by-product to flow down the tube, full return is not being realized. When industry becomes more aware of this situation, steps will most assuredly be taken to plug up that type of subtle dollar drain-off.

The implication of the above is this: We in anaerobic marketing and design have the primary responsibility to educate industry in the subject of anaerobic treatment. We must make clear to management the economic and environmental advantages, and the attendant return on investment.

Accomplishing this educational process will be to the benefit of both industry - and ourselves.

MONSANTO PERSPECTIVE ON ANAEROBIC TREATABILITY OF INDUSTRIAL WASTES

William E. Gledhill, Edward G. Valines, and Michael F. Weishaar

Monsanto Company
St. Louis, Missouri

ABSTRACT

This paper summarizes results of various research projects concerning anaerobic treatment of different Monsanto industrial waste waters. Experiences from studies that attempted to apply anaerobic technology to process effluents have resulted in no workable solutions to date. Problems of component inhibition, high total dissolved solids and low COD streams have been responsible for the lack of success. Even with our limited success anaerobic technology is still considered a good alternative to waste activated sludge and worth continued investigation.

INTRODUCTION

Traditionally, anaerobic treatment has not worked well with wastewaters other than those from municipal or food processing (distillery, beverage, vegetable) sources. Industrial wastewaters from the chemical industry are often complex, containing a wide variety organics unrelated to the carbohydrate structures found in the municipal or food processing wastes. Though many industrial chemicals are amenable to anaerobic metabolism (7) process wastes usually do not contain a single chemical component. Complex process intermediates, polymers and toxicants often are encountered which defy any biological treatment. In addition, much of the previous work on evaluation of anaerobic treatment of industrial wastes has not been conducted with knowledge of anaerobic metabolism and has overemphasized yield on methane.

The ever changing regulatory environment demands an ongoing evaluation of existing waste water treatment schemes for process wastes which have historically been treated using conventional methods. Anaerobic technology (1,3,4, 7,8) offers distinct advantages over its aerobic counterpart : less sludge production economic operation, low nutrient requirements, high microbial biomass and potential for energy recovery. This paper presents a sampling of anaerobic waste water treatability studies conducted by Monsanto and includes a review of the anaerobic screening methodology used for environmental fate studies.

ANAEROBIC SCREENING TECHNIQUE

Environmental fate studies to assess the persistence of chemicals often involve anaerobic biodegradability. Monsanto has helped refine the anaerobic biodegradability procedure developed by Young (5) and reported by Shelton and Tiedje (6). In this procedure a chemical or waste stream containing at least 5 mg carbon is exposed in a 160 ml serum bottle to 100 ml of a chemically defined medium containing 1 mg/l resazurin, an oxidation/reduction indicator and 10% (v/v) anaerobic sludge from a local domestic sewage treatment plant. The chemical's biodegradability or toxicity is assessed by measurement of gas production using a Unimeasure/80 pressure transducer with a P-8 bellows (Unimeasure, Inc.). Theoretical gas production from 5 mg carbon is 10.5 ml at 35 deg. C and atmospheric pressure. The relative amounts of carbon dioxide and methane are determined by reaction stoichiometry with the Boswell equation (9). Alternatively, head space gases can be measured via gas chromatography or, for radiochemical analysis, by sparging and trapping radiolabelled gases. We have also used this procedure to assess biodegradability in an anaerobic sediment/water system simulating a lagoon situation.

The anaerobic biodegradability of sodium polyglyoxylate (SPG), a potential new detergent phosphate replacement, exemplifies the use of this technique. SPG's persistence is dependant on its rate of hydrolysis to sodium glyoxylate (2). Although SPG is classified as readily biodegradable in aerobic

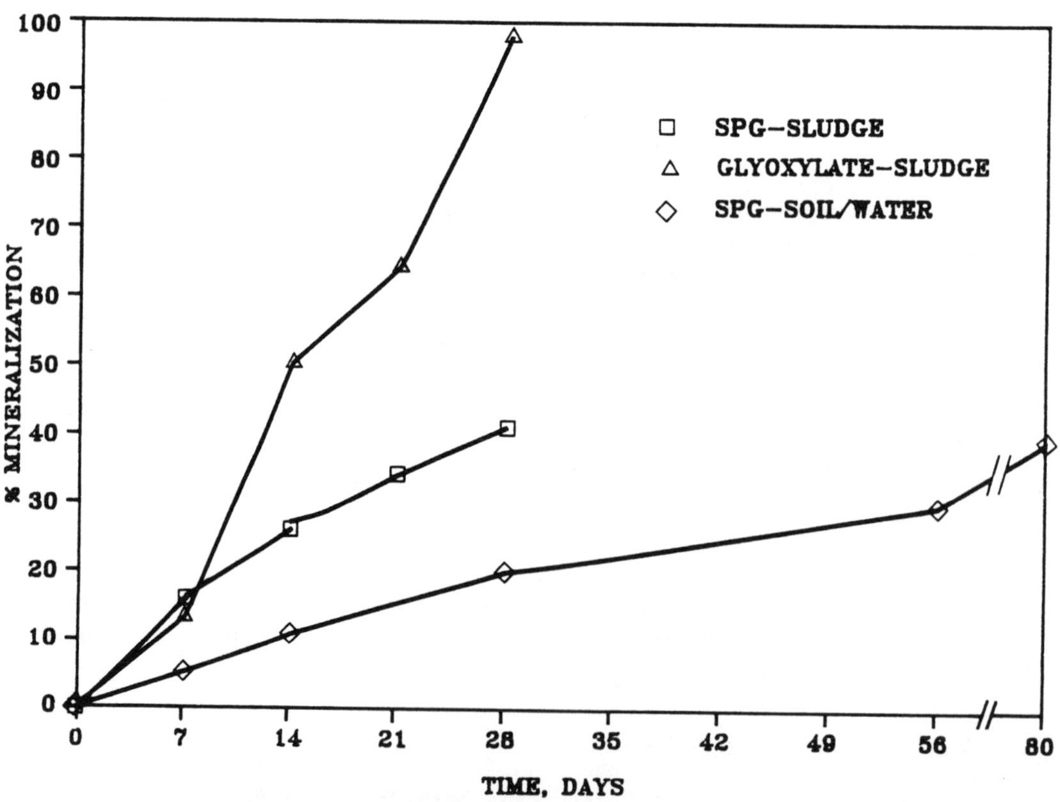

FIG. 1. Anaerobic Biodegradation Of SPG

biodegradation screening tests its rate of hydrolysis is not fast enough to actually degrade during primary or secondary treatment. The material forms an insoluble calcium salt which settles out in primary and secondary clarifiers and, subsequently, finds its way to anaerobic digestors. SPG's hydrolysis, and consequently biodegradability, is facilitated by the longer SRTs in anaerobic digestors. Figure 1 shows that SPG and sodium glyoxylate are biodegradable under anaerobic conditions. Also shown is SPG's biodegradation in an anaerobic sediment/water system as assessed by monitoring radiolabelled carbon dioxide and methane. Studies also have shown that SPG is not toxic to anaerobic sludge at concentrations anticipated in the digestor. These screening studies have verified in actual laboratory scale digestors. As the process for manufacturing plant wastes.

MONSANTO ANAEROBIC TREATABILITY CASE STUDIES

Monsanto has contracted several anaerobic treatability studies with various consulting firms. Our experience is not extensive and we have found no uniquely suitable waste stream for scale up of an anaerobic process. The examples I will present are relatively brief attempts to evaluate the anaerobic susceptibility of various waste streams. We have gained significant experience from these studies and even though none have been successful we are not abandoning our hopes on this technology.

One of our earliest studies involved a fiber intermediate waste stream containing a soluble COD of 41,000 mg/l, BOD5 of 10,000 mg/l, Kjeldahl nitrogen of 6,100 mg/l and ammonia nitrogen of 680 mg/l. The experimental design involved dilutions of the wastes in anaerobic sludge in syringes. Syringes were analyzed for gas production at 70 and 102 days. COD removal was 43%, ammonia nitrogen increased by 36-53 % and net methane production was zero. Inhibitory compounds were also present in the waste stream; however, effects were reduced with dilution.

Approximately one year later a highly pretreated version of the same waste water, containing 3,500 mg/l COD, 857 mg/l BOD5, 660 mg/l Kjeldahl nitrogen and 160 mg/l ammonia nitrogen, was retested to examine if deamination could occur under anaerobic conditions. In spite of the pretreatment anaerobic treatment was only moderately effective at deamination of this waste stream.

Results of another treatability are summarized in Table 1. Batch studies indicated that this acetic acid process waste was susceptible to anaerobic treatment. Continuous flow studies conducted over three month period confirmed that an anaerobic process would be an attractive process for this particular waste. High organic loading was achieved (1-2 g SOC/l/d), good SOC removal occurred (>60%), comparatively low sludge volumes were generated and high methane efficiencies were achieved (>95%). In the final analysis the process was not implemented since the waste was treatable aerobically and the economics would not justify construction of the digestor (~$2M). Perhaps in the future if other waste streams at this site are amenable to anaerobic treatment the acetic acid stream will be incorporated into the process.

Table 1. Anaerobic treatability of an acetic acid process waste.[1]

	Batch[2]				Continuous Flow[3]	
Reactor	1	2	3	4	1	2
Loading						
%	0.4	0.8	1.7	3.3	-	-
g SOC/l/d	-	-	-	-	1.5	1.4
Infl.SOC, mg/l	490	1010	2000	3600	1463	1436
Effl.SOC, mg/l	150	270	500	3540	566	499
% Removal	69	73	75	2	61	65
Gas						
l/g SOC Removed	1.29	1.16	1.24	-	0.81	1.04
% Methane	-	-	-	-	96	95

[1] Waste characteristics : pH 2.6; NH3 - N, 590 mg/l; Kjeldahl - N, 5550 mg/; TDS, 72,900 mg/l; COD, 564 g/l; SOC, 245 g/l; acetic acid, 20 - 25%; misc. organics, 5%.
[2] 500 ml bottles
[3] 12 l Anhybrid and Anbiof Reactors.

An additional study was initiated to examine the anaerobic treatability of a pesticide waste stream and specific components which were resistant to aerobic treatment. Overall anaerobic treatability was not as successful in terms of BOD/COD removal as was the aerobic process. Table 2 summarizes the results. The specific pesticide, which is regulated under NPDES permit for our aerobic system, was not removed in anaerobic treatment. However, process by products and a degradation intermediate, which are not effectively removed in aerobic treatment, showed some signs of removal in this study. The study, however, lasted only 33 days and suffered from low influent TOC, low MLSS and short HRT's. No definitive conclusions, apart from determining that the waste was not toxic to the anaerobic microorganisms, could be based on this study. Consequently, a more in depth study was undertaken.

Table 2. Anaerobic screening of a pesticide waste stream. [1]

Parameter	Influent	Effluent	% Removal
TOC, mg/l	1512	652	57
BOD 5	1868	916	51
Volatile acids, mg/l	1196	487	59
Pesticide, mg/l	166	224[2]	neg.
Pesticide Interm., mg/l	312	110	65
Pest. by prod.#1, mg/l	125	57	55
Pest. by prod.#2, mg/l	1788	929	48

[1] Operating conditions : 5 gal. glass carboys; anaerobic lagoon sludge as seed; Temp., 32 deg. C; MLSS 2131 mg/l; MLVSS, 761 mg/l; HRT, 6-16 d.
[2] Increased due to desorption from sludge inoculum.

Examination of a similar pesticide waste stream was subsequently carried out (Table 3). As in the previous study MLSS and HRT were low., consequently; organic loadings were relatively high (0.2 -0.3 g SOC/l/d or almost 1 g COD/l/d). Sludge had poor settleability characteristics so that recycle to the digestor was not possible. Thus, sludge residence times were less than 10 d which is borderline in terms of methanogen washout. However, in spite of these limitations gas production was good and averaged 500 ml/d at 80% methane. SOC removal was judged to be adequate for the HRT and SRT used. The specific pesticide was not removed. Because of the problems encountered in this study it was recommended to examine an anaerobic system , such as a fixed film reactor, permitting both higher organic loadings and biomass. These studies have not been initiated to date.

Table 3. Anaerobic treatability of a pesticide waste stream.

	Reactor 1			Reactor 2		
	Influent	Effluent	% Removal	Influent	Effluent	% Removal
SOC,mg/l	3433	2644	23.0	2227	1479	33.6
Pest.,mg/l	416	402	3.4	209	269	0

Waste characterization : COD, 22,800 mg/l; SOC, 13,200 mg/l; TDS, 10%; pH 7.3. Operating conditions : 3l continuous flow reactors, Reactor 1 HRT = 11.1 d; Reactor 2 HRT = 6.5 d; gas production 500 ml/d; MLSS 2100 - 3500 mg/l.

The final case involves one of the more extensive studies to date. The waste water was a combination of a fiber intermediate, specialty chemical and agricultural chemical process stream with the following characteristics: TOC 5,000 mg/l, COD 11,000 mg/l, ammonia nitrogen 720 mg/l, organic nitrogen 880 mg/l and sulfate 1,075 mg/l. The study involved anaerobic batch testing, up flow sludge bed flow through reactors, down flow completely mixed reactors with solids recycle and a large upflow two zone hybrid reactor. Reactors were all operated at organic loadings of less than 0.05 g COD/ g MLVSS. Results in the up flow and down flow reactors indicated 20% TOC reduction and 60% organic nitrogen conversion to ammonia. In the dual zone reactor 35% TOC reduction and 80% organic nitrogen conversion to ammonia were achieved. Gas production from the waste stream was negligible. Thus, even though deamination was somewhat successful COD removal and methane production were inadequate.

SUMMARY AND LESSONS LEARNED

The specific examples above point to significant problems Monsanto has encountered in its efforts to apply anaerobic treatment. These problems have not yet been surmounted; consequently, Monsanto has to date not applied anaerobic technology to a full scale operation. We have, however, gained significant experience in the course of these investigations. Our studies have shown us that just as in aerobic treatment there are chemicals or conditions which inhibit the biological process. Examples such as excessive ammonia production from deamination, excessive TDS interfering with normal sludge growth and recalcitrant organics toxic to microorganisms have been encountered. In the cases where removal of specific compounds were examined anaerobic treatment was either ineffective or results were not definitive. In some cases methane production was low. A key point is that industry, while certainly interested in methane production for cost savings, would not discount anaerobic treatment technologies because of low gas production. If removal of specific toxic and/or regulated components can be demonstrated, we would definitely consider using it.

Our lack of success is probably attributable to the fact that we were looking at wastes in the "hard to treat" category. Also we have not made a concerted effort to apply anaerobic technology to our general menu of waste streams. There may indeed be instances where waste streams now disposed of by aerobic treatment can more cost effectively be treated anaerobically. Also, more in depth studies involving a combination of chemical and biological (anaerobic) treatment may provide suitable treatment for hard to treat wastes now disposed of via non biological means. Despite the limited success of our studies Monsanto considers anaerobic treatment as an attractive alternative to aerobic waste water treatment processes and we will continue research in this area.

REFERENCES

1. Forster, C.G. 1984. The potential of anaerobic digestion for treating liquid wastes. Water Poll. Control pp. 484-490.

2. Gledhill, W.E. 1987. Environmental degradation of sodium polyglyoxylate, a non persistent metal sequestrant. Development in Ind. Microbiol. 26: In press.

3. Harvey, L.M. and J.C. Rubiano. 1983. The anaerobic approach: A better answer. Ind. Wastes, Mar./Apr. pp.22-25.

4. Oleszkiewicz, J.A. and M. Olthof. 1982. Anaerobic treatment of food industry wastewaters. Food Technol., June, pp. 78-82.

5. Owen, W.F.,D.C. Stuckey, J.B. Healy, Jr., L.Y. Young and P.L. McCarty. 1979. Bioassay for monoritoring biochemical methane potential and anaerobic toxicity. Water Res. 13 : 485-492.

6. Shelton, D.R. and J.M. Tiedje. 1984. General method for determining anaerobic biodegradation potential. Appl. Environ. Microbiol. 47: 850-857.

7. Speece, R.E. 1983. Anaerobic biotechnology for industrial wastewater treatment. Environ. Sci. Technol. 17 : 416A-426A.

8. Switzenbaum, M.S. 1983. Anaerobic treatment of wastewater : Recent developments. ASM News 49 : 532-536.

9. Tarvin, D. and A. M. Buswell. 1934. The methane fermentation of organic acids and carbohydrates. J. Amer. Chem. Soc. 56 : 1751-1755.

INDUSTRIAL-SCALE ANAEROBIC TREATMENT OF YEAST FERMENTATION WASTEWATER

Denise A. Johnston

BIOTHANE Corporation
Camden, New Jersey

ABSTRACT

The BIOTHANE high-rate anaerobic process has been in continuous operation at two bakers yeast manufacturing facilities for a cumulative total of nearly four years. The anaerobic system is used as a primary step in the treatment of wastewater originating from several different molasses fermentation streams. At one facility the anaerobic effluent is discharged directly to the municipality, while at the other facility the anaerobic effluent is first polished for ammonia removal in a secondary activated sludge system and then discharged to the municipality. Excellent operating results have been attained at both facilities. BOD purification efficiencies of 85% to 95% at volumetric loading rates of 10 to 14 Kg COD per cubic meter of digester volume per day are routinely achieved. This paper briefly summarizes some of the design and operating parameters for each facility.

INTRODUCTION

Pretreatment of wastewater through the use of anaerobic technology in order to resolve discharge problems is rapidly gaining acceptance within the food processing industry. A good example of this is the situation which exists at two major bakers yeast manufacturing facilities in the United States. At Dixie Yeast Corporation in Gastonia, North Carolina, a high-rate anaerobic pretreatment process is employed to reduce BOD discharge to the municipality for regulatory compliance reasons. At Busch Industrial Products (BIP) in Old Bridge, New Jersey, the same type of high-rate anaerobic system is in use because of the economic benefits to be gained from reduction of municipal surcharges and in-plant utilization of the methane gas byproduct. The anaerobic systems at these two plants have each been in continuous operation for approximately two years.

THE BIOTHANE SYSTEM

The BIOTHANE process is an outgrowth of years of developmental work which took place at Dutch universities during the late 60's and early 70's. The process embodies upflow anaerobic sludge blanket (UASB) technology through the use of specially designed patented digester topworks which work efficiently to retain biomass within the digester and allow very high loading rates (short hydraulic retention times) to occur. This allows for greater space efficiency and more precise system control, both features of importance to most industries. The first industrial-scale BIOTHANE system was installed at a beet sugar plant in Europe in 1978, and there are now more than 50 such plants in operation or under construction throughout the world, 9 of which are in the United States.

TREATMENT OF BAKERS YEAST WASTEWATER

A pilot program was carried out at BIP during 1982 using a BIOTHANE six cubic meter digester. Loading rates in excess of 10 Kg COD per cubic meter of digester volume per day were achieved with BOD removal efficiencies averaging close to 85%. The study had several focal points including the proper internal flow pattern necessary to optimize treatment of the high-strength waste resulting from selected streams (COD greater than 25000 mg/l) and the stability of treatment efficiency while operating at somewhat colder temperatures (approximately 88 to 90°F) than the "normal" mesophylic operating range of 92 to 98°F. The successful results achieved during the pilot program resulted in the decision by BIP to proceed with design and installation of a full-scale system the following year.

Dixie Yeast Corporation personnel took a somewhat different approach to their treatment problem and made the decision to pretreat the entire (composite) factory stream rather than selecting specific high-strength streams. A pilot program utilizing a BIOTHANE 10-liter continuous flow bench reactor was carried out during the summer of 1983 in conjunction with an aerobic polishing study (conducted by the engineering firm CH2M Hill) designed to determine ammonia conversion efficiency. BOD removal efficiencies of about 90% were realized at loading rates similar to those achieved during the BIP study completed previously. A similar ratio of BOD to COD purification potential was found during both studies, indicating the presence of similar levels of non biodegradable substances in the wastewater which show up as COD in the standard testing methods. The construction of a full-scale system at Dixie started early in 1984, and, by October of 1984, the system was started into operation.

SYSTEM PERFORMANCE OVERVIEW

Some of the key design parameters for each bakers yeast facility are listed in **Table 1** along with the operating "average" for each parameter. The relatively weaker strength of the Dixie composite wastewater stream compared to the selected streams at BIP is apparent from the Table, as is the fact that in both cases the actual wastewater COD concentration coming into the anaerobic system is about 25% weaker than projected in the design stage. This is compensated by the higher-than-design wastewater flows in each case which, when taken together, result in a close match between design and operating COD Volumetric Loading rates. It can be seen that the BOD purification efficiency achieved at Dixie on the weaker stream is somewhat higher than that achieved at BIP, although the COD removal efficiencies in both plants are similar. Since Dixie has a much longer BOD sampling history than does BIP, this difference may be due to measurement error.

The methane fraction of the biogas is higher at Dixie than at BIP, but the methane yield (volume of methane obtained from conversion of COD) is similar at both plants. Finally, it can be seen from the Table that in all important respects the operating performance of each plant equals or exceeds that projected in the design stage.

Table 1. BIOTHANE SYSTEM PERFORMANCE SUMMARY FOR TREATMENT OF YEAST FERMENTATION WASTEWATER

	DIXIE YEAST		BUSCH INDUSTRIAL PRODUCTS	
	DESIGN	OPERATING	DESIGN	OPERATING
DIGESTER VOLUME (m^3)	1750	1750	5000	5000
FLOW (MGD)	0.23	0.32	0.40	0.50
INFLUENT COD (mg/l)	21000	15000	33500	25000
COD REMOVAL (%)	62	65	60	65
BOD REMOVAL (%)	85	95	80	85
CH_4 PURITY (%)	70	75	70	65
COD VOL. LOAD. (Kg/m^3/d)	10.9	10.7	10.3	9.5
BIOGAS PRODUCTION (m^3/d)	5068	5097	14600	16000
METHANE YIELD (m^3/Kg COD)	0.30	0.32	0.33	0.34

DAILY SYSTEM PERFORMANCE

In order to illustrate the daily performance of the BIOTHANE system with exposure to variations in factory effluent characterisitics, COD and BOD data have been plotted in **Figure 1** for one year of operation at Dixie Yeast Corporation. The period selected is April 1985 through March 1986, approximately 0.5 to 1.5 years following system start up. Influent and effluent soluble COD data are shown on the top plot, while influent and effluent BOD data are shown in the middle plot. The bottom plot shows the COD volumetric loading data. It is striking that the consistency of the effluent BOD appears to be independent of rather substantial daily variations in COD volumetric loading data which, in turn, is a reflection of the combination of wastewater strength and hydraulic flow through the system. It can be seen that the system was exposed to volumetric loadings ranging from 4 to 16 Kg/m^3/d during the period. The individual point nature of the plots (rather than a continuous curve) is due to the sampling of these parameters on a several times per week basis. In actual fact, the system was operated every day during this period with no down time.

DISCUSSION AND SUMMARY

The operating data presented here show a great similarity in performance between the anaerobic pretreatment plants at the two bakers yeast facilities. In both cases the wastewater is derived from the fermentation of molasses, although at one plant only high strength streams have been selected for treatment, while at the other the composite effluent is treated. A progression from pilot study to full-scale plant design took place in both cases, and, in both cases, the full-scale plant performance has proven to be as good as or better than that projected from the pilot data. Each full-scale treatment plant has been in continuous operation for about two years, and the year-long data summary shown in Figure 1 for the Dixie plant is indicative of the kind of stability which can be achieved with a system of this type. Both manufacturing facilities have gained economic advantages through the decision to pretreat their wastestreams anaerobically. Similar potential benefits await other industrial users.

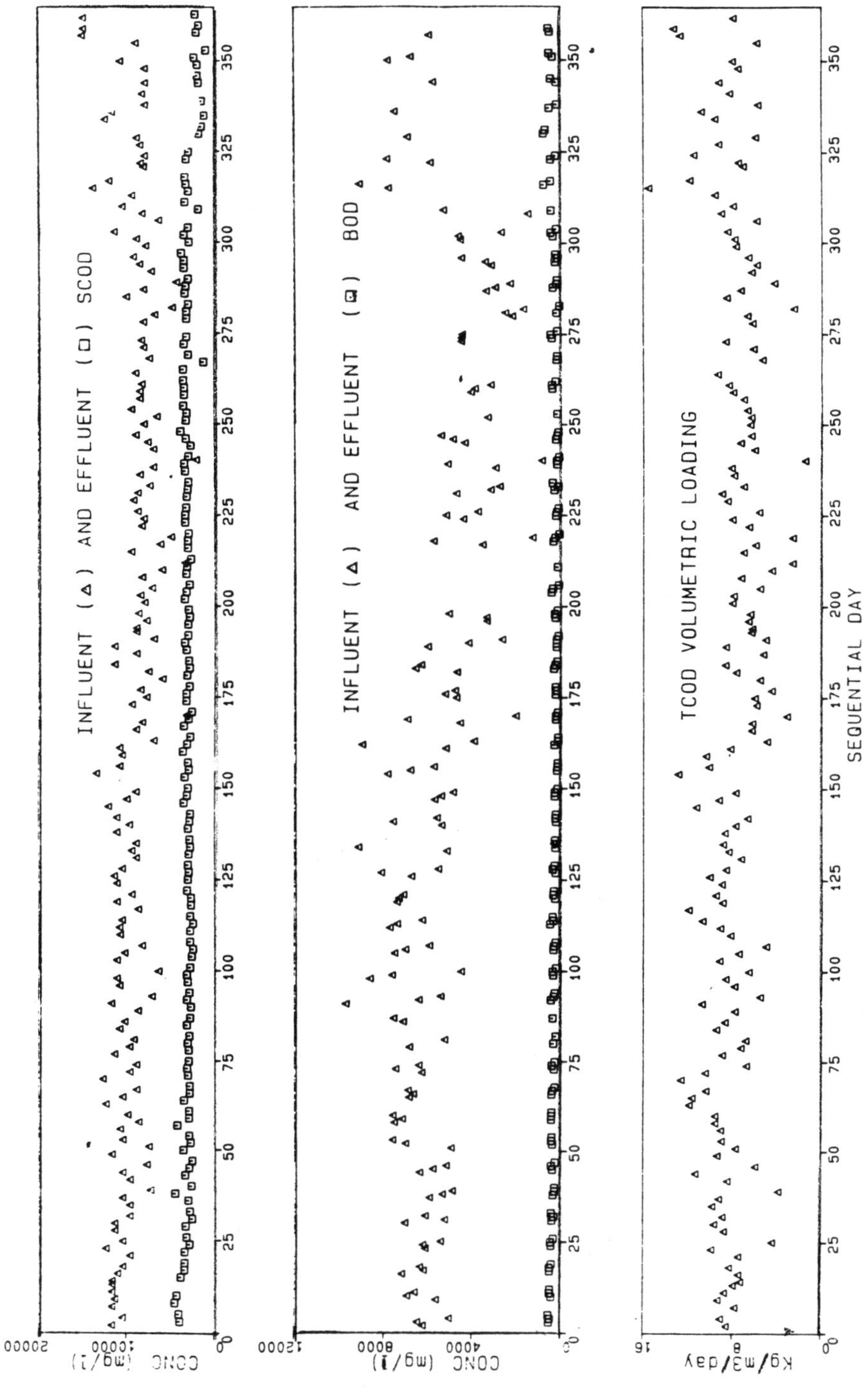

FIG. 1 BIOTHANE SYSTEM PERFORMANCE AT DIXIE YEAST FOR PEROID APRIL 1985 THRU MARCH 1986

ANAEROBIC SEQUENCING BATCH REACTOR TREATMENT OF COAL CONVERSION WASTEWATERS

James P. Earley and Lloyd H. Ketchum, Jr.

Department of Civil Engineering
University of Notre Dame
Notre Dame, Indiana

ABSTRACT

This work is a continuing laboratory study to attempt development of both an operating strategy and a basis of design for Anaerobic Sequencing Batch Reactor (AnSBR) biological treatment of coal conversion wastewaters. The project is in the early stages and the results reported are based on only several months of investigation. Support is being provided by the U.S. Department of Energy, Pittsburgh Energy Technology Center (Grant No. DE-FG22-85PC80512). The laboratory studies are being conducted in three different size reactors. The smallest are 150-mL serum bottles that are being used to screen individual coal conversion wastewater constituents. Several constituents have been shown to be degraded under anaerobic conditions. Six 2-L AnSBR's are being operated to acclimate anaerobic sludges (i.e., organisms) to mixtures of these constituents. Finally, two automatically operated, 4-L AnSBR are being used to treat a synthetic coal conversion wastewater currently consisting of phenol, (the only carbon source) and vitamins and minerals.

ANAEROBIC SEQUENCING BATCH REACTORS

These are anaerobic biological reactors operated in a fill and draw mode. The reactor cycle begins with FILL, a period of time when raw wastewater is pumped into the reactor which contains the biomass from previous cycles. Mixing is provided to promote contact between the biomass and influent organics. At the end of REACT, mixing is discontinued and settling of the biomass is provided, SETTLE. The treated supernatant is removed during DRAW. IDLE is a period between the end of DRAW and the beginning of the next FILL period.

SERUM BOTTLE STUDIES

The screening tests are conducted in triplicate. The three serum bottles are each filled with an anaerobic sludge (these have been collected from several different sources, including a mixture of sludges collected from 10 different municipal anaerobic digesters), a vitamin and mineral solution and one of the coal conversion wastewater constituents. The gas production is monitored, and compared to serum bottle controls that contain only the vitamin and mineral solution and anaerobic sludge. Inhibition is noted whenever gas production is less than the control, and degradation is noted whenever gas production is higher and compares to the theoretical gas production for that constituent and initial concentration. Table 1 shows these results recorded after either 60 or 90 days of observation.

Table 1. Serum Bottle Screening Results

Constituent	No. of Triplicates	Initial Conc. Range, mg/L	Days to complete Degradation, Range	Possible Inhib. at Conc., mg/L
Phenol	8	200-1000	10-40	1000
o-Cresol	6	200-500	none in 90 days	200
m-Cresol	7	100-500	20-70	
p-Cresol	5	200-500	12-19	
Aniline	7	100-250	none in 90 days	
2,3-Xylenol	7	100-250	none in 90 days	200
2,4-Xylenol	2	100-250	none in 90 days	250
2,6-Xylenol	2	100-250	none in 90 days	250
3,4-Xylenol	7	100-250	none in 90 days	200
3,5-Xylenol	7	100-250	none in 90 days	200
Catechol	4	200-1000	45-66	1000
Resorcinol	4	200-1000	24-50	
Hexanoic acid	2	250-500	less than 1 day	
Benzoic acid	3	200-500	5-7	
Pyridine	2	150-250	28	
Guaiacol	2	150-250	14-20	
4-Methylcatechol	2	150-250	45-48	
2,3,5-Trimethylphenol	3	100-1000	none after 60 days	250
2,4,6-Trimethylphenol	2	100-250	none after 60 days	250
3,4,5-Trimethylphenol	2	100-250	none after 60 days	250

ACCLIMATION STUDIES

Six 2-L AnSBR's are being monitored. These have been started using different sludge source mixtures and startup policies. The reactors are currently being fed a synthetic coal conversion wastewater (CCW) prepared as follows:

Constituent	Conc., mg/L	Equivalent TOC, mg/L
Phenol	4000	3060
o-Cresol	500	385
m-Cresol	500	385
p-Cresol	500	385
Aniline	50	39
3,4-Xylenol	50	39
3,5-Xylenol	50	39
Acetic acid	400	160
Propionic acid	125	61
n-Butyric acid	50	28
Nutrients and vitamins	-	-
Total TOC		4581

The liquid volume of each reactor is 1.5L. The reactors are held at 36°C and fed 38 to 100 mL of feed each week day. Magnetic stirrers are used to keep the reactors mixed. Two of the reactors are being fed a mixture of CCW and glucose, three others are being fed CCW diluted with distilled water and one is being fed full-strength CCW. Table 2 shows the current operating conditions of each reactor.

Table 2. Acclimation Studies Operating Policy

Operating Parameter	Reactor Designation					
	R1	R2	R3	R4	R5	R6
Influent TOC, g/L	5.1	4.9	1.1	2.3	2.3	4.6
Weekday Feed, mL	100	75	100	75	75	38
Source of TOC						
% CCW	22	46	100	100	100	100
% Glucose	78	54				
H_2O Dilution	none	none	1:4	1:2	1:2	none
Reactor MLVSS, g/L	8.7	6.3	2.5	2.9	3.8	4.1

In each reactor, all of the following are being degraded by the end of each cycle: glucose, phenol, acetic acid, propionic acid, and n-butryic acid. There has been no evidence of degradation of aniline or the xylenols. The cresols are being degraded, but at different rates and after different periods of time. These results are most interesting and can be seen on Figure 1. Figure 1 shows the effluent concentration of the cresols in each reactor during the 3.5 months of the study. For comparison, the conservative element concentration (i.e., the effluent concentration if no reactions took place) has also been shown (i.e., the upper curve). It appears that the presence of glucose inhances o-cresol degradation, but inhibits degradation of m-cresol and p-cresol degradation (i.e., compare R1 and R2 with glucose to the others).

4-L AnSBR STUDIES

Two reactors are being operated automatically as follows:

- FILL - 3 hours; 180 mL feed containing 4000 mg/L phenol and minerals and vitamins.
- REACT - 16 hours; with mixing; 3.6 L total volume; pH = 7, SS = 5 g/L, and VSS = 3.5 g/L.
- SETTLE - 4 hours.
- DRAW - 0.5 hours; 180 mL is removed which results in a 20 day hydraulic retention time.
- IDLE - 0.5 hours; which completes a 24 hour cycle.

After only one and one-half month of operation, phenol degradation is nearly complete, no pH adjusting chemicals are added and effluent suspended solids concentration are improving and have reached 50 mg/L.

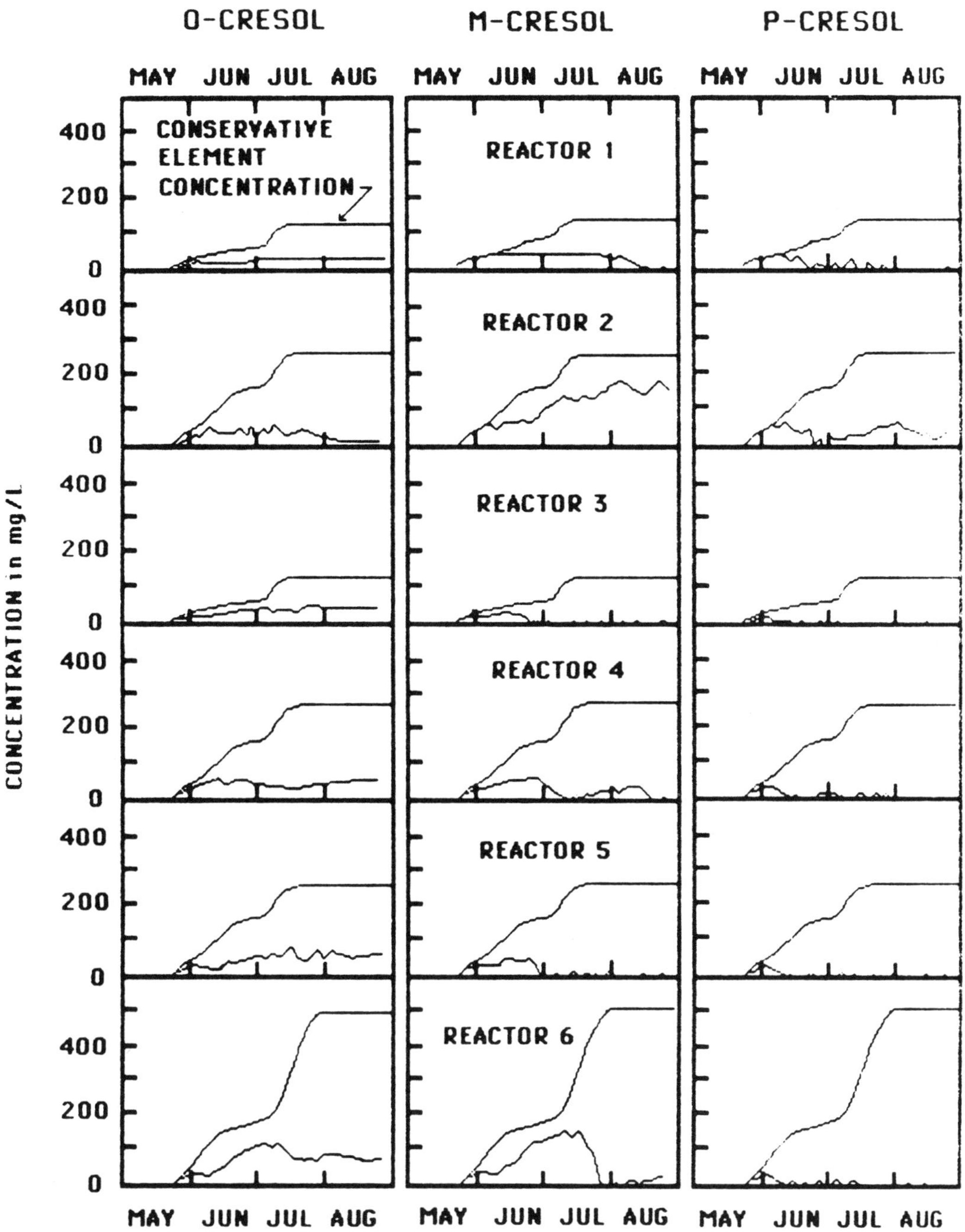

Figure 1. Cresol Degradation in AnSBRs

POTENTIAL FOR ANAEROBIC TREATMENT OF HIGH SULFUR WASTEWATER IN A UNIQUE UPFLOW–FIXED FILM–SUSPENDED GROWTH REACTOR

L. Syd Love

Sydlo, Inc.
Mississauga, Ontario, Canada

ABSTRACT:

Anaerobic degradation is ideally suited for the pre treatment of high strength industrial effluents. However, many, particularly those from the Pulp and Paper Industry, may contain substantial amounts of sulfur. This will result in sulfide toxicity and inhibit anaerobic degradation.

An obvious solution would be to reduce the soluble sulfide concentration within the reactor, to a point below the "threshold limit". This can be done in the Sydlo Anaerobic Reactor, by rapid removal of biogas.

A unique design feature, of the Sydlo Anaerobic Reactor, makes it possible to remove large volumes of supernatant including biogas containing hydrogen sulfide (H_2S), from the reactor at two intermediate points. This recycle flow, which can be very high, is directed to external gas separators, where the biogas (and H_2S) are removed. The supernatant, less biogas (and H_2S), is then returned to the reactor.

Our poster session will describe the Sydlo Anaerobic Reactor and show how it can be successfully used for the treatment of high sulfur wastewaters.

UNDESIRABLE EFFECTS OF H_2S:

The production of H_2S during anaerobic degradation is undesirable because of:

a. It is odorous - "rotten eggs".
b. It is corrosive.
c. It is toxic - "sulfide toxicity".*
d. The sulfate bacteria utilize organic matter that otherwise would be converted to methane gas.

*Sulfide toxicity poses a serious problem to anaerobic degradation.

SULFIDE TOXICITY:

If the sulfide concentration in an anaerobic reactor exceeds 200 mg/L (as S) all gas production ceases.

There appears to be a linear relationship between gas production and the sulfide concentration within the reactor.

Sulfides are less inhibitory to the acid forming bacteria than to methane forming bacteria.

PRESENT DAY SOLUTION (SULFIDE TOXICITY)

Add ferric chloride - precipitate inert ferric sulfide. Process works well but results in high operating cost.

a. Ferric chloride is expensive.
b. Increases sludge disposal cost.
c. Precludes possibility of sulfur recovery.

GAS LAWS:

Both Boyles Law and Henry's Law state that the H_2S in the gas phase of an anaerobic reactor is in equalibrium with the H_2S dissolved in the liquid phase.

Therefore if H_2S is rapidly removed from the gas phase the solubility of H_2S in the liquid phase will be greatly reduced.

ALTERNATIVE SOLUTION (SULFIDE TOXICITY)

In the Sydlo Anaerobic Reactor, shown in Figure No.1, large volumes of supernatant are withdrawn and directed to an external gas separator.

Biogas, including the H_2S, is removed by the gas separator. The supernatant, less the biogas and H_2S, is then recycled back to the Sydlo reactor.

This procedure will rapidly remove H_2S from the gas phase and greatly reduce the solubility of H_2S in the liquid phase.

This supernatant recycle rate can be adjusted, to greatly reduce, or completely eliminate, sulfide toxicity.

ADVANTAGES OF HIGH RECYCLE RATE:

The basic advantage is a substantial reduction in operating cost:

a. Elimination of ferric chloride addition.
b. Lower sludge disposal cost.
c. Possibility of sulfur recovery from H_2S gas.

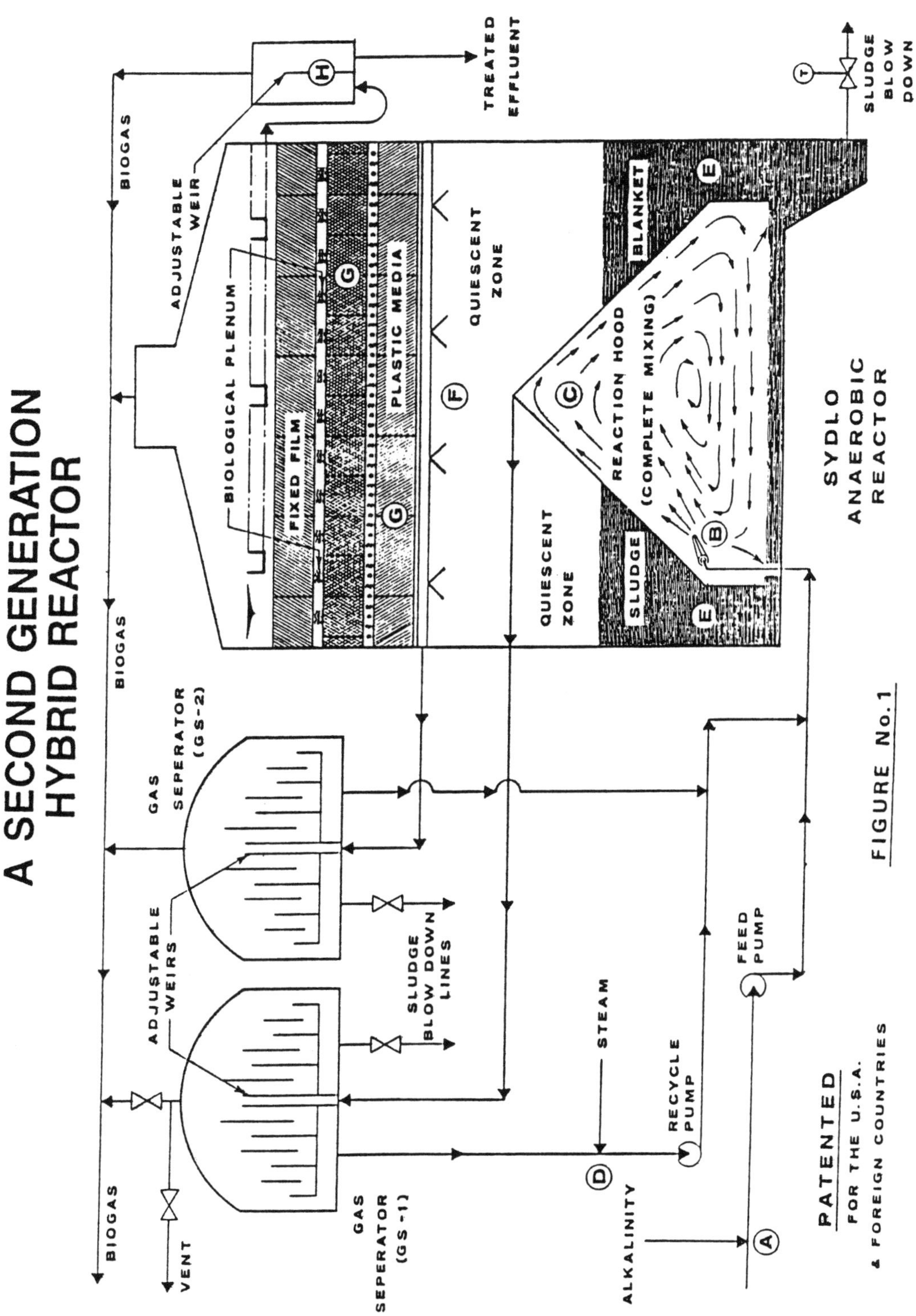
A SECOND GENERATION
HYBRID REACTOR
BIOGAS
VENT
GAS SEPERATOR (GS-1)
GAS SEPERATOR (GS-2)
ADJUSTABLE WEIRS
SLUDGE BLOW DOWN LINES
ALKALINITY
STEAM
RECYCLE PUMP
FEED PUMP
BIOLOGICAL PLENUM
ADJUSTABLE WEIR
FIXED FILM
PLASTIC MEDIA
QUIESCENT ZONE
REACTION HOOD
(COMPLETE MIXING)
SLUDGE
BLANKET
TREATED EFFLUENT
SLUDGE BLOW DOWN
SYDLO ANAEROBIC REACTOR
PATENTED
FOR THE U.S.A.
& FOREIGN COUNTRIES

FIGURE No.1

THE HIPERION PROCESS: AN ADVANCED METHOD OF HYDROGEN SULFIDE REMOVAL FROM BIOGAS

Jackson Yu

Ultrasystems Engineers and Constructors, Inc.
Irvine, California

Abstract

Among the bacteria responsible for anaerobic digestion of wastewater is desulfavibrio desulfurican. Sulfur-bearing compounds in wastewater streams are generally converted to hydrogen sulfide by these bacteria. Hydrogen sulfide thus produced must be removed from the biogas in order to mitigate both environmental and operational problems associated with biogas usage. In the past, the removal of hydrogen sulfide in the presence of carbon dioxide has been a difficult or expensive task. Ultrasystems Engineers and Constructors, Incorporated (Ultrasystems) has helped to develop an economical catalytic process to convert the hydrogen sulfide present in biogas to elemental sulfur. The catalyst, based on 1,4 naphthaquinone, is regenerated with air for reuse on a continuing basis. Commercial units of the Hiperion process are operational, and additional units are currently under design.

Inherent to the anaerobic digestion of wastewater containing sulfate or other sulfur-bearing substrates is the generation of hydrogen sulfide along with the biogas. The treatment of biogas for removal of sulfur compounds becomes more and more important as both government regulations restricting sulfur emissions become tighter and the effect of corrosion from biogas usage creates maintenance and operational problems. In most cases, better than 99 percent sulfur removal is necessary. This means that the hydrogen sulfide level in the effluent gases must often be reduced to 10 ppm or less to compensate for the traces of organic sulfur compounds, carbonyl sulfide, and other sulfurous materials which are more difficult to remove. Assuming an inlet hydrogen sulfide concentration of two volume percent, or 20,000 ppm, the reduction down to 10 ppm represents 99.95 percent removal.

Historically, hydrogen sulfide, with or without carbon dioxide, has been concentrated by liquid scrubbing with amines followed by a Claus unit to convert the hydrogen sulfide to sulfur. The requirement for very low effluent gas sulfur levels mentioned above means that even a Claus unit must be followed by a tailgas cleanup process. In other words, three processing steps (hydrogen sulfide scrubbing, Claus conversion, and tailgas cleanup) are necessary for total and effective desulfurization.

If the gas to be treated is lean in H_2S (less than 20 volume percent) or the amount of recoverable sulfur is relatively small, as in the case of biogas (0.1 - 10 tons/day), the Claus route is not feasible. The use of caustic scrubbing or ferric chloride addition has not been cost effective, as these methods convert a gaseous environmental problem to another form which may require further solution.

Ultrasystems has developed an effective alternative catalytic oxidation/reduction process that converts hydrogen sulfide directly to elemental sulfur without the use of heavy metals or other hazardous chemicals.

Direct oxidation processes to reduce hydrogen sulfide from gases have been in use for over 100 years. Several of the earlier processes utilized iron oxide suspension in alkaline solution. With names like Burkheiser, Ferrox, Gluud, and Manchester, these sulfur removal techniques were applied mainly to town gas from coal in Europe and the United Kingdom. Today, these operations are no longer competitive with direct oxidation processes utilizing liquid or soluble catalytic agents.

It has been recognized that the quinone family of chemicals, which has high redox potentials, could be effective in converting hydrogen sulfide to sulfur. However, the direct use of a quinone catalyst without other intermediate reactions did not materialize until the Takahax process, which uses 1,4 napthaquinone-disulfonic acid as the catalytic agent, was developed by Dr. N. Hasebe of Japan. Today, there are over 150 facilities in Japan which use the Takahax process for sulfur reduction from coke oven gases, chemical plants, and anaerobic digesters.

The Hiperion process, also developed by Dr. Hasebe, is an improved version of the Takahax process. The major difference is the formulation and behavior of the catalyst. Hiperion makes use of a naphthaquinone chelate, which permits a high oxidation rate resulting in a significant reduction in reaction residence time compared with the Takahax process. This, in turn, results in smaller equipment and lower costs. In other respects, the Hiperion and Takahax processes are similar.

The principal reaction in the removal of hydrogen sulfide involves the redox transfer of the napthaquinone chelate and H_2S to form naphthahydroquinone chelate and sulfur. The naphthahydroquinone chelate is regenerated by air oxidation to reform naphthaquinone chelate and water. The absorption and oxidation reactions are shown below:

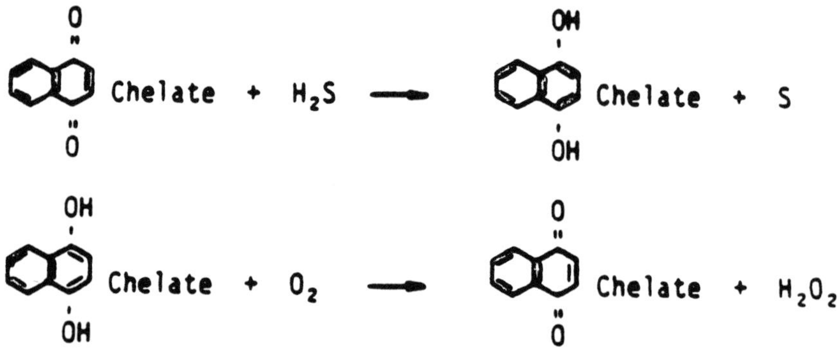

$$H_2O_2 + H_2S \rightarrow 2H_2O + S$$

$$2H_2S + O_2 \rightarrow 2H_2O + 2S$$

In the Hiperion process, the reactions are carried at near ambient gas temperatures. In a conventional Claus unit, reaction temperature on the order of 400°F is required to make the reaction proceed at a significant rate. The higher temperature limits equilibrium hydrogen sulfide conversion in Claus units to about 97 percent, whereas conversions in the Hiperion unit are essentially 100 percent complete.

With its capacity for rapid and essentially quantitative removal of hydrogen sulfide, the Hiperion process is ideally suited for the elimination of sulfur from a variety of gas streams (e.g., refineries, coke ovens, natural gas plants, pulp and paper mills, waste treatment facilities, chemical facilities, and geothermal units). Illustrated in Figure 1 is a simplified flow diagram of the Hiperion process. The sour gas containing sulfurous compounds is fed into the bottom of the absorber column where the gas is contacted with an aqueous solution of the naphthaquinone as it cascades downward in the absorber. Sweet gas is removed from the top of the absorber. The reduced solution is transferred from the bottom of the absorber to the top of the regeneration tower. Air is introduced through a sparger in the bottom of this tower to regenerate the spent catalyst. A slip stream from the bottom of the regenerator is fed to a filter where the sulfur is removed. The regenerated naphthaquinone chelate solution is then recycled to the absorption tower. Both towers have several packed sections. The patented packing, "windy balls," is designed for fluidization to minimize potential sulfur deposition, provide for minimum holdup of sulfur, and to reduce overall pressure drop.

The high redox potential of the Hiperion catalyst provides for high chemical reactivity with hydrogen sulfide to form elemental sulfur. This is translated into a high throughput rate. The Hiperion catalyst is also completely selective to hydrogen sulfide, even in an acid gas with high CO_2 concentration. Wide pH swings are not experienced; therefore,

Figure 1. Flow Diagram of the Hiperion Process

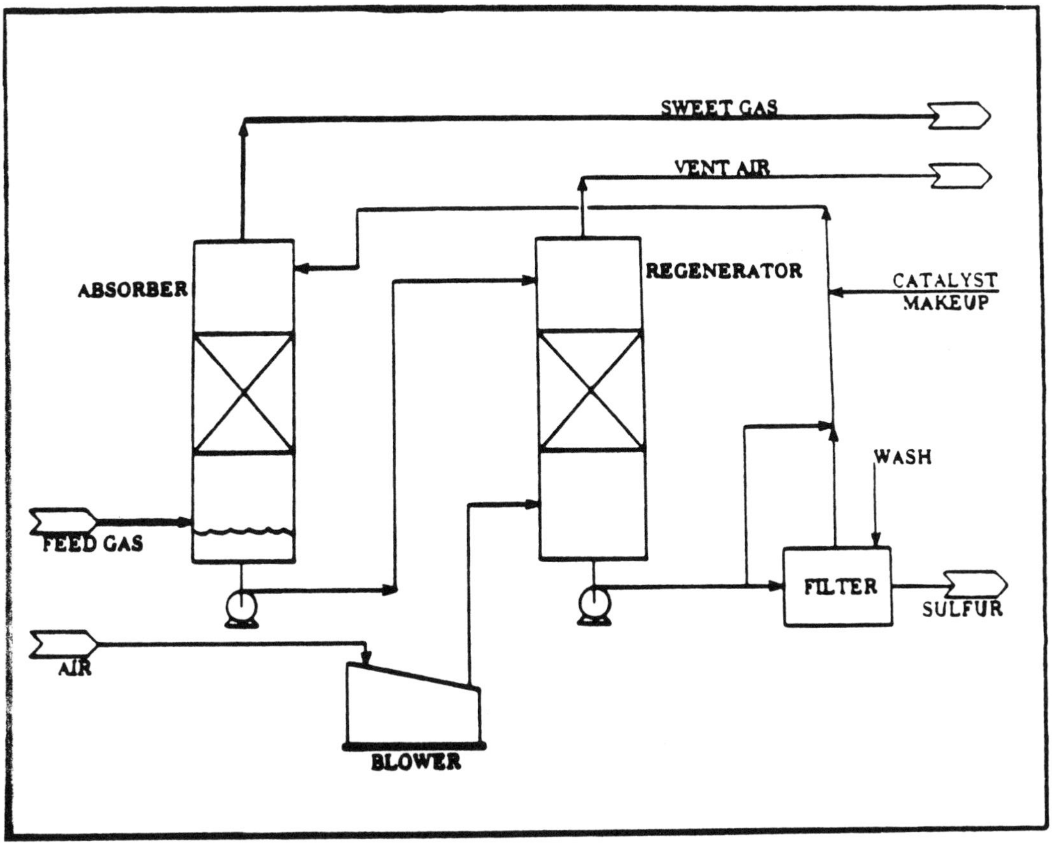

the use of alkalai for pH buffering is minimized. The process is not sensitive to the operating pressure and temperature of the system. Where low or high feed gas temperature is dictated, solvents other than water may be used. A wide range of solvents including alcohols, glycols, and glycerins can be accommodated by the process.

Formulation of the Hiperion catalyst does not require the use of hazardous additives. The catalyst solution is, therefore, considered nonhazardous and nonflammable. The catalyst is not consumed in the reactions. It is regenerated each time with air, and its performance is not deteriorated with time. Therefore, the only source of catalyst loss in the Hiperion process is attributable to entrainment and physical adhesion of the catalyst on the sulfur cake. Reclamation through cake washing minimizes catalyst loss. The Hiperion catalyst is very stable, and it is not subject to aging, self-decomposition, or biodegration.

In addition to Hiperion's technical merits, the process is also cost effective. Most processes that remove hydrogen sulfide under high CO_2 partial pressure consume alkaline material to maintain pH. The Hiperion catalyst is formulated to minimize CO_2 absorption and, as a result, pH changes are also minimized. Any CO_2 that is physically absorbed in solution is generally desorbed during catalyst regeneration. With essentially no chemical addition, other than to compensate for physical losses of the catalyst, total chemical costs are minimal.

Because of the high reactivity of the catalyst, the Hiperion process offers a low liquid circulation rate. This low circulation rate results in reduced energy consumption. There is no byproduct formation in the Hiperion process. A continuous bleed stream to prevent an excess buildup of impurities in the catalyst solution is not required. Because the Hiperion process does not use any toxic additives in the catalyst, spillage and waste can be safely treated by conventional means. This results in significant savings in waste disposal costs and no requirement for specialized treatment facilities. Its small space requirements and simple process scheme allow the Hiperion plant to be easily integrated into a waste treatment plant or other facilities.

All of the above-mentioned factors result in an overall simple desulfurization process characterized by a lower solution circulation rate and a more efficient, yet compact plant. The Hiperion process provides savings in capital and operating costs.

DORR-OLIVER'S EXPERIENCE WITH ANAEROBIC TREATMENT OF INDUSTRIAL WASTEWATERS

Paul M. Sutton and Timothy W. Bezler

Dorr-Oliver, Inc.
Stamford, Connecticut

ANITRON™ PROCESS DESCRIPTION

The Anitron system (Figure 1) is a highly efficient anaerobic wastewater treatment process which utilizes a fluidized bed reactor. Within the reactor, a fixed-film of microbial growth (supported growth) occurs on the media (usually sand), which are hydraulically supported as a fluidized bed by the incoming wastewater and recycled effluent. Wastewaters with BOD_5 levels of 2000 mg/l or more, such as found in the food, beverage, and pulp and paper industries, are candidates for treatment with this technology. Tables 1, 2 and 3 and Figure 2 summarize performance results of the Anitron system with key industrial wastewaters.

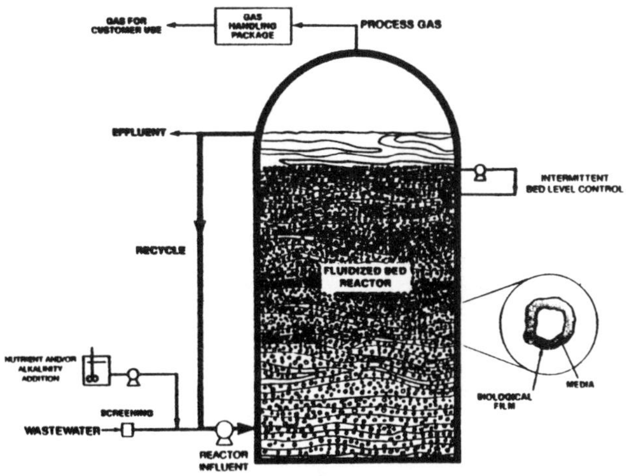

Figure 1. Anitron Process Schematic

Table 1. Correlation Between Pilot- and Full-Scale Performance on Soy Processing Wastewater (Grain Processing Corporation)

	Pilot Scale	Full-Scale
Mean Volumetric Loading (kg COD/m^3-day)	13.0	15.6
Mean Feed Values (mg/l)		
COD	10,914	10,556
BOD_5	6,556	7,960
Removal Efficiency (%)		
COD	-	76
BOD_5	83	83

Following is a summary of the performance of the Anitron system at various volumetric loadings. Note the excellent COD removal efficiencies at 24 kg COD/m^3-day (92%) and 35 kg COD/m^3-day (86%), and that the system was still functioning acceptably at 102 kg COD/m^3-day.

Table 2. Anitron Performance at High Volumetric Loadings on Pulp and Paper Wastewater (EPA-Funded Study, Meehoopany, PA, 1983)

	Test Period (days)		
	97-124	168-179	193-198
Loading (kg COD/m^3-day)	24	35	102
Detention, days	0.52	0.29	0.10
% COD Removal	92	86	69
% BOD Removal	93	81	74
Gas			
m^3/kg COD_R	0.48	0.50	0.46
% CH_4	69	69	62
Influent mg/l			
COD	14,500	13,900	13,300
BOD	10,800	11,100	11,400
VOA	9,410	9,290	8,600
SS	73	29	106
Effluent mg/l			
COD	1,150	1,900	4,180
BOD	740	1,840	3,650
VOA	820	1,120	2,390
SS	140	124	147

Table 3. Two-Phase Anitron System: Corn Processing Wastewater American Maize-Products Company

	Corn Processing Wastewater		Corn Processing Wastewater with Waste Activated Sludge	
Volumetric loading kg COD/m^3-day	16.0	18.7	20.2	25.4
Feed values, mg/l				
COD	8840	9310	10193	8430
BOD_5	5770	5196	6211	5306
Effluent values, mg/l				
Soluble COD	772	1072	1077	1163
Soluble BOD_5	397	504	673	903
Removal, %				
COD	91.5	88.5	89.4	86.2
BOD_5	93.1	91.7	89.2	83.0

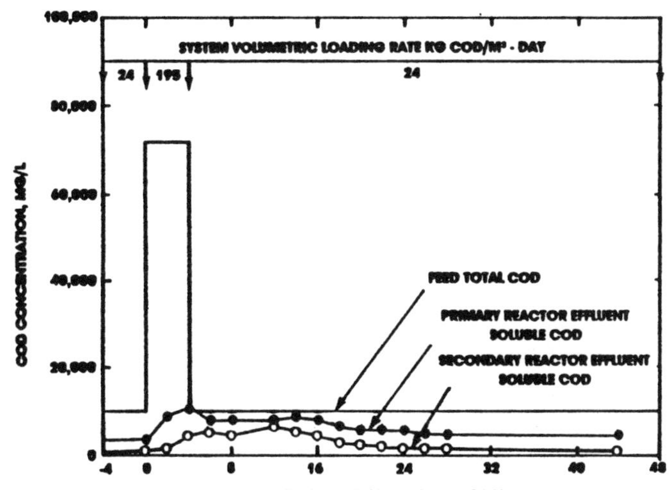

Figure 2. Response to Shock Loading Conditions (American Maize)

MARS™ PROCESS DESCRIPTION

The Membrane Anaerobic Reactor System (MARS) wastewater treatment process (Figure 3) represents the ultimate in biological treatment technology. A suspended growth - or contact - reactor is coupled with ultrafiltration membrane modules which serve to completely retain system biomass and provide an effluent that is essentially free of suspended solids. Candidate wastewaters include industrial streams with COD greater than 20,000 mg/l) such as cheese whey or permeate) and low flow wastewaters where conventional aerobic technology is being considered. Table 4 indicates MARS performance results with ultra high-strength waste streams.

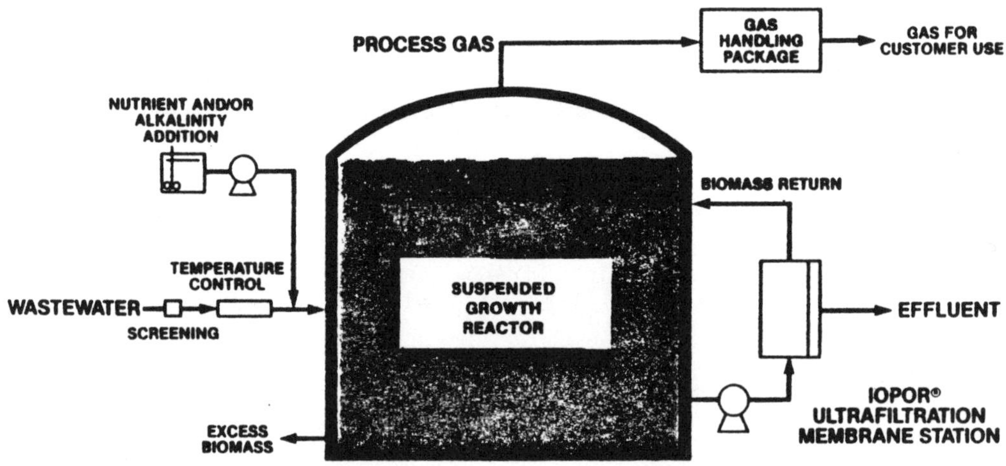

Figure 3. Anaerobic MARS Process Schematic

Table 4. MARS Performance with Various Waste Streams

Waste, Location, Date	Mean Influent Values (mg/l)			Mean Effluent Values (mg/l)		
	COD	BOD_5	TSS	COD	BOD_5	TSS
Wheat Starch Ontario, 1985	11,240	4,200	2,267	194	30	< 10
Wheat Starch Minnesota, 1983-84	35,175	15,463	13,300	270	74	< 10
Sweet Whey Permeate Minnesota, 1984	59,790	28,417	512	305	-	< 10
Acid Whey Permeate New York, 1983-84	55,107	23,502	107	454	27	< 10

HYAN PROCESS TREATS HIGH STRENGTH WASTEWATER AND GENERATES USABLE ENERGY

David W. Hein and George V. Crawford

G.S. Processes, Inc.
Scottsdale, Arizona

The Thermal Conditioning Process at the Lakeview Water Pollution Control Plant produces a high strength waste stream that constitutes 20% of the total load of organic pollutants on the Aerobic Biological Treatment System. The Lakeview facility is a municipally owned wastewater treatment plant located in Mississauga, Ontario, Canada. In order to expand the capacity of the plant, two options existed: either expand the aerobic treatment facility or reduce the biological load by treating the high strength wastes separately. The HYAN hybrid anaerobic process was developed to effectively treat this high strength waste and to generate a continuous and reliable supply of gas energy.

The new process design has successfully overcome the problems encountered by other anaerobic systems attempting to treat high strength wastes. In particular high loading rates can now be maintained without causing either pluggage of the reactors with solids or loss of active biomass from the system.

The HYAN Process is a cost effective method of treating a variety of high strength waste streams. Energy recovery, reduced operating costs, low capital costs for equipment and minimal land requirements combine to give the HYAN System a demonstrated payback that is superior to alternative systems.

The Problem

Thermal conditioning liquors (TCL) are generated at the Lakeview plant as a byproduct of the sludge dewatering operations. As part of the thermal conditioning process, biological cells in the sludge are hydrolyzed and oxidized. A significant portion of the particulate solids matter is converted to dissolved solids. After the conditioned solids have been separated from the liquid by settling and dewatering, the liquid phase contains high concentrations of pollutants in the form of Biological Oxygen Demand (BOD) and Chemical Oxygen Demand (COD). This polluted liquid stream must undergo further treatment before it can be discharged to receiving waters. Traditional aerobic biological treatment of this stream has always been expensive.

Innovative Technology

Because of the nature of the waste, and because of its obvious potential for anaerobic treatment, the Region of Peel and the Ontario Ministry of the Environment commissioned Gore & Storrie Limited in 1978 to conduct a laboratory scale investigation into anaerobic biological treatment of the thermal conditioning liquors generated at Lakeview. Subsequent studies tested a variety of anaerobic processes and configurations including sludge blanket, fluidized bed and filter reactors. Upflow and downflow filters were studied and random and modular type filter media were investigated. In all, three separate studies were conducted by Gore & Storrie Limited and one by Environment Canada. As well, Gore & Storrie engineers visited the United States and Germany where the anaerobic treatment of thermal conditioning liquor was being attempted.

None of the anaerobic processes available at the time were found to be suitable. To treat the Lakeview TCL, Gore & Storrie developed a unique hybrid configuration combining anaerobic suspended growth and fixed film filter technology. Such a design configuration had never before been researched or used elsewhere.

In January 1983 the hybrid anaerobic reactor configuration was constructed at laboratory scale by Gore & Storrie in the firm's laboratory. That reactor received Lakeview TCL as feed. Throughout the design phase and continuing through 1986, removals of 70% of soluble COD and 88% of soluble BOD were achieved in the hybrid reactor. Reactor loadings were above 6.3 kilograms of COD per cubic metre per day at a total reactor hydraulic retention time of 38 hours. These loadings are high, relative to other anaerobic processes treating similar wastes. As a result of the laboratory scale testing, the HYAN process was adopted, implemented and commissioned at full scale.

Full Scale Application

In 1985, the full scale hybrid anaerobic process was put into operation to treat the byproduct liquors. The HYAN process has consistently removed 80 to 85% of the BOD pollutant load from the waste stream. At this rate, plant treatment costs will be reduced $150,000 per year. Additionally, the HYAN produces a continuous supply of fuel gas worth an estimated $310,000 per year. The HYAN process has turned an expensive wastewater treatment problem into a cost-saving energy-producing opportunity.

The Lakeview facility was constructed by modifying two existing 27 metre diameter digesters. The liquid volume of each reactor is 2910 cubic metres. The $1.5 million cost included all related construction and control building rehabilitation. The facilities were commissioned in April 1985 and achieved full design loading by June. Through much of 1985 and 1986, the reactors have treated all the thermal conditioning liquors produced by the plant.

Operating Results

The facility has consistently achieved 72% COD and 80% BOD reductions, while producing between 10,000 and 14,000 cubic metres of fuel gas per day. The gas is transmitted to the adjacent Lakeview Bioenergy System complex, increasing the energy utilization and production of that facility.

The reduced load to the aerobic facility has lowered the overall electrical treatment costs. In addition, the methane produced has replaced most of the natural gas requirements of the Thermal Conditioning Process.

The reduced loading on the Aerobic System has deferred major capital expenditures for new tankage and aeration systems until increased sewage flows justify additional plant capacity. The HYAN system has also reduced the quantity of solids requiring treatment and disposal, by efficiently converting the organic pollutants to gas.

Investment Payback

The new hybrid anaerobic process has transformed what was a complex, expensive wastewater treatment problem into a highly-effective opportunity to reduce treatment costs while generating substantial amounts of fuel gas of great economic value.

Annual Electrical Power Costs to improve waste stream to discharge quality	
- by Aerobic System only	$190,000.00
- by Aerobic System and HYAN	$ 40,000.00
Power Cost Savings	$150,000.00
Value of Methane Fuel Produced	$310,000.00
Annual Cost Savings	$460,000.00
HYAN Construction Costs	$1,500,000.00
Payback Period	39 months

The alternative to the HYAN System would have been to expand the Aerobic Biological Treatment System at an estimated capital cost of 3.9 million dollars.

Widespread Applicability

The HYAN process was developed specifically for the treatment of the Lakeview TCL where solids growth, accumulation and precipitation during anaerobic treatment were of critical importance and concern. The process is equally applicable for many other high strength wastes. Such wastes include landfill leachates, pharmaceutical, dairy, food processing, textile, petrochemical and pulp and paper wastes. The treatment of many of these wastes has already been demonstrated.

Anaerobic processes are prone to high accumulations of inert solids in their fixed film filter zones when treating high strength wastes. Most designs require frequent shut-downs and the use of difficult cleaning procedures to reduce short-circuiting and to maintain performance. The HYAN reactor controls such accumulations by design, keeping its filter media zone free of unwanted solids.

The HYAN concept is also being effectively applied where the upflow sludge blanket process has had difficulty maintaining biomass in the reactor. The HYAN design is proving to be more efficient in those cases by not requiring a separate settling tank to capture and return biomass to the reactor.

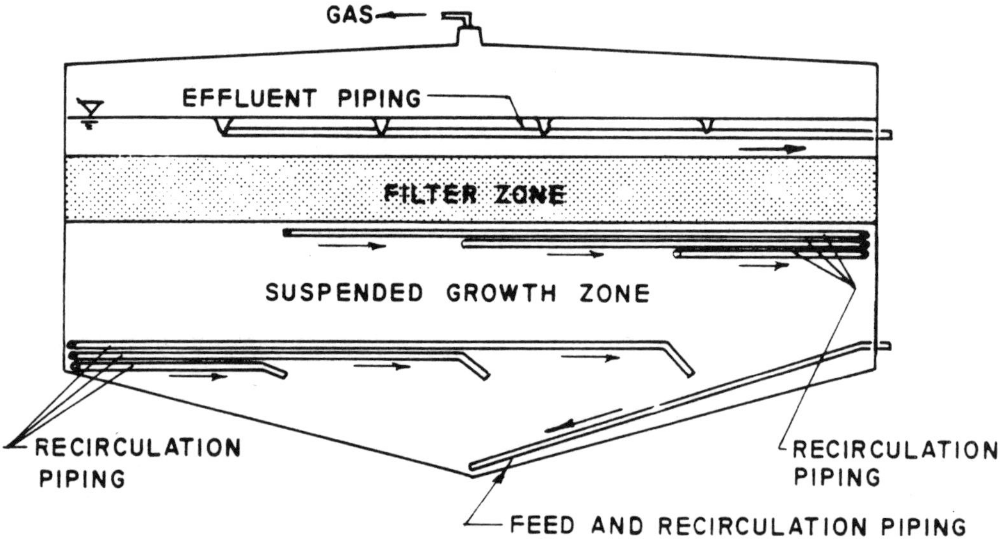

- THE HYAN REACTOR -

Biological growth and pollutant conversion takes place in the suspended growth zone, producing gas and biomass solids. Equalized feed distribution and mixing of the suspended growth are achieved by recirculation of the lower zone contents. Recirculation flow rate adjustments can be made without causing solids to rise in the filter zone where undesirable accumulation would occur.

The upper reactor is based on anaerobic filter technology to take advantage of its high performance and resistance to shock loads. Gas rising from the lower zone provides vertical mixing between the two reactor zones. The agitating action of the rising gas dislodges and returns solids to the lower zone, keeping the filter media clean.

GS Processes Inc. offers the HYAN Process technology and design under license in the United States.

David Hein and George Crawford are the Construction Manager and Technical Director, respectively, of GS Processes Inc. Mr. Hein is currently responsible for the design, procurement and construction of all HYAN projects in North America. Mr. Crawford has worked almost exclusively on the development and application of anaerobic processes for the last seven years, and was the project manager and process designer for the Lakeview project.

BIOSULFIX: AN ANAEROBIC TREATMENT PROCESS FOR HIGH SULFATE WASTESTREAMS AND SLUDGES

Gregory R. Gillespie and Meint Olthof

Duncan, Lagnese, and Associates, Inc.
Pittsburgh, Pennsylvania

ABSTRACT

Over the past several years, attention has been focused on anaerobic treatment of high strength industrial waste streams and sludges. In this research, the presence of sulfates, sulfites, and sulfides has been considered a nuisance. In an anaerobic reactor, the sulfate reducing bacteria (SRB) reduce sulfates to sulfides that can create a toxic environment for the methane forming bacteria (MFB). The sulfides produced end up in the bio-gas formed by the anaerobic reactor. The sulfides present cause corrosion problems and odors. The work presented represents DLA's development of an anaerobic process called "BIOSULFIX" that degrades the organics and forms valuable by-products, mainly sodium bisulfide (NaSH).

BACKGROUND INFORMATION

SRB's are found in nature. SRB's propagate in river and estuary sediments where the presence of organics and sulfates allow the SRB to grow. They can also be found in municipal sewage. There has been considerable research conducted on the microbiology of these organisms. Originally it was thought that they were only able to oxidize organics to acetic acid. With time, however, it was discovered that there are SRB's that can oxidize the organics all the way to CO_2 and H_2O. The types of organics that can be totally oxidized by the SRB's include lactic, propionic, acetic and amino acids; sugars; alcohols; cheese whey; organic containing waste streams from industry and municipal sewage sludge.

The SRB's are able to outcompete the MFB's in most situations where they can propagate side by side. There are only a few waste streams that contain enough sulfates to allow the SRB's to oxidize all organics in the waste. Therefore, in most cases, one will have the hydrogen sulfide in conjunction with the methane (provided that the sulfide concentration is below toxicity limits). Literature generally agrees that 200 mg/l hydrogen sulfide may result in toxicity problems. By maintaining a low hydrogen sulfide concentration in the liquid, it is possible to have both SRB's and MFB's working side by side.

RECOVERY OF SULFUR FROM FGD AND MUNICIPAL SEWAGE SLUDGES AND WASTEWATER

The City of Lakeland, Florida has the unique position to own both a municipal waste treatment plant and a power plant. The municipal treatment plant needs to be upgraded, while the power plant generates flue gas desulfurization (FGD) sludge. A laboratory feasibility study was conducted to determine the feasibility of mixing the undigested sludge from the municipal sewage plant with the FGD sludge. The objective was to first digest the organics in the sludge so that the residual organic sludge can be used for land disposal, while at the same time the sulfur in the gas is recovered as sodium bisulfide that can be sold to various industries including the pulp and paper industry.

The results showed that 44% of the VSS can be removed, which is similar to methane digestion. The hydrogen sulfide formed was about 0.5 grams per liter of reactor per day. With about 47 grams per liter of TSS in the reactor, this equates to a specific sulfur production rate of 0.01 grams S per gram TSS per day. Although during the initial stages of the reactor performance methane gas was formed, after several weeks of steady-state operation, no methane gas was formed during the rest of the study. The sulfate reducing bacteria had outcompeted the methane forming bacteria.

The results of the experiment were used to make an economic analysis of applying "BIOSULFIX" in a full scale facility. It showed that the major new unit operations needed were the FGD slurry handling system, hydrogen sulfide absorption system, a classifier, and a thickener. The hydrogen sulfide absorber is a newly developed process that will selectively scrub the hydrogen sulfide out of the gas stream. In those applications where there is a mixture of methane, carbon dioxide, and hydrogen sulfide, it is important to use the caustic for scrubbing as efficiently as possible and absorb primarily hydrogen sulfide. A separate process was developed for this selective absorption. In the digestion of the sludges, the FGD sludge is converted to limestone, therefore there is a separation process required to separate the digested organic sludge from the calcium carbonate. The classifier is used to do this. The calcium carbonate can then be returned to the power plant for scrubbing, while the thickened digested organic sludge can be used for land disposal. This classification step is optional since the value of the limestone is very small in the overall economics. The major appeal for this process, as far as economics is concerned, is the generation of sodium bisulfide which sells for about 30¢ a pound as a 44% solution. The pulp and paper industry is a large consumer of this chemical in their pulping process. Other consumers are the tannery industry, the organic chemical industry, and the dye stuff manufacturing industry. The value of the by-product for the City of Lakeland (70,000 inhabitants) is about $1.1 MM/year.

SULFATE CONTAINING WASTE FROM SPECIALTY CHEMICAL MANUFACTURING

For a number of years, DLA conducted a laboratory scale treatability and pilot studies on high strength waste streams from a specialty chemical manufacturing facility. The organics present in the waste were alcohols, chelating agents, and fatty acids. The tests were conducted in continuous fed reactors.

Subsequent to the bench scale treatability studies for the specialty chemical manufacturer, a larger scale pilot study was conducted on this waste. For six months, the bacteria mass was allowed to increase and this resulted in a higher allowable organic loading to the reactors. As with any biological process, the true performance should be measured based on a pound of VSS basis. The kinetics based on a volume of bioreactor is not as meaningful as the reaction rates based on the amount of biomass. In a well operated anaerobic bioreactor, it is possible to maintain a volatile suspended solids of 10 to 20 grams per liter.

The off-gases from this bioreactor were treated with caustic to selectively absorb the hydrogen sulfide. In this case, the sodium bisulfide formed could be used as a raw material in another specialty organic chemical plant.

The overall performance of the "BIOSULFIX" system was marginal with respect to SOC/BOD removal. BOD removal was 50% on the pilot plant versus 85% in the lab scale reactors. The poor performance of the pilot plant was due to the variable nature of the waste. The wastewater from the specialty chemical manufacturing facility was high in dissolved solids. Another contributing factor was the length of the pilot study. The small scale laboratory reactors were operated over a period of two years. The pilot plant did not have adequate time to develop its full biomass potential in the six-month study.

SUMMARY

The results of the studies indicate that "BIOSULFIX" is a viable alternative for sulfate wastes. By using anaerobic treatment technology to treat industrial wastes, industries may be able to minimize their waste disposal operation costs.

Development work utilizing the "BIOSULFIX" process is continuing on both the laboratory and pilot scale. Investigations are being conducted utilizing two-stage digestion, an acetogenic and sulfidegenic reactor, to evaluate the impact of phase separation on the performance of the "BIOSULFIX" process. Pilot work has commenced on applying "BIOSULFIX" to chrome tanning wastes. Both of these studies are on-going and will be featured in future presentations by DLA on this subject.

REFERENCES

1. BURGESS, S.G. AND L.B. WOOD
Pilot Plant Studies in Production of Sulfur From Sulfate Enriched Sewage Sludge.
J.Sci. Food Agric., 12, April, 1961.

2. LOVLEY, D.R. et al
Kinetic Analysis of Competition Between Sulfate Reducers and Methanogens for Hydrogen in Sediments Applied and Env. Microbiology, June, 1982, p.1373.

3. MIDDLETON, A.C. et al
Kinetics of Microbial Sulfate Reduction.
Journal Water Pollution Control Federation (1977), July, p.1659.

4. OLESZKIEWICZ, J.A. et al
Anaerobic Treatment of High Sulfate Wastes.
University of B.C., Conference on New Direction and Research in Waste Treatment, At University of B.C., June 22, 1985.

5. OBARSKY, B.J. et al
Sulfur Removal of Polysulfide Rubber Manufacturing Wastewaters by Anaerobic Treatment.
Purdue University Proceedings, 33rd Conference, p.402.

6. PIPES, W.O. JR.
Sludge Digestion By Sulfate Reducing Bacteria.
Purdue University Proceeding, 1960, p.308.

7. RABOLINI, F. et al
Biological Treatment of Acid Mine Water.
NTIS - PB 213 930, July, 1971.

8. SADANA, J.C. and A.V. MOREY
Microbiological Production of Sulfide From Gypsum.
J. Sci.Industr.Res. Vol.21C, May, 1962.

9. SCHONHEIT, P. et al
Kinetic Mechanism for the Ability of Sulfate Reducers to Out Compete Methanogens For Acetate.
Arch.Microbiol.(1982) 132:285-288.

10. SMITH, J.R. AND A.C. MIDDLETON
Microbial Sulfate Reduction For Reclamation of Sulfate Wastes.
Koppers Company, February 1980.

THE CELROBIC PROCESS–PERFORMANCE STABILITY

A.M. Sobkowicz

Badger Engineers, Inc.
Cambridge, Massachusetts

ABSTRACT

The Celrobic* high-rate anaerobic treatment process was developed by the Celanese Chemical Company, and is currently being used commercially in nine installations. The upflow random packed-bed configuration accounts for the reliability and stability of the process. Its long-term continuous performance relies on the ability to measure and control the quantity of solids that remain in the reactor.

INTRODUCTION

The Celrobic high-rate anaerobic wastewater treatment process has been used for over eight years. Over this period it has been shown to perform stably over a wide range of waste types and loadings. It has also proven to be faster and more reliable than other anaerobic systems in starting up and in achieving its design loading and removal.

The first commercial Celrobic unit, located in Vernon, Texas, started up on the waste from a guar bean processing plant in 1977.

After two-and-a-half years of accumulation, solids plugged the reactor. It was shut down, cleaned, and restarted. This incident provided the impetus for a research program into methods for measuring and controlling the accumulation of solids in the Celrobic reactor.

BIOMASS CONTROL DEVELOPMENT

Celanese Chemical Company identified tracer chemicals which are extremely bio-refractory, but do not interact in any way with the microorganism in the reactor. Using one of these tracers the volume within the reactor that is not occupied by solids can be measured accurately, thereby assisting in biomass control.

Concurrently, methods for stripping accumulated solids from the reactor internals in order to prevent plugging were developed in the Celanese commercial units, which by that time had increased to three (described in Tables 1-3). Operating performances for the three Celanese systems are shown in Figures 1-3. The research program proved to be successful; after the 1980 shutdown of the Vernon reactor, no Celrobic reactor has had to be shut down because of accumulated solids.

Along with the COD removal efficiency, the void volumes measured using a tracer chemical for the Bishop, Texas and Pampa, Texas systems are shown in Figures 2 and 3. For the Bishop system, the removal efficiency

*Celrobic is a service mark of the Celanese Corporation.

tracks the void volume closely, illustrating that a relatively high void volume seems to promote better removal. On the other hand, a high inventory of biomass (low void volume) is thought to produce a more stable system for difficult wastes (i.e., wastes that are highly variable or that contain toxins). At Pampa, a lower void volume is maintained, with the expectation that some of the solids might be digested under these conditions. A satisfactory removal still results.

COMMERCIAL EXPERIENCE

Table 4 lists other installations where the Celrobic process is used commercially. Its ability to treat a wide variety of wastes as diverse as those from the food processing, fermentation, and petrochemical industries demonstrates the versatility of this technology. The long-term reliability of the process has been amply demonstrated in the continued good performance of the early installations.

Table 1

Vernon Plant

Location:	Vernon, Texas (150mi NW Dallas)
Major Feedstocks:	Guar Beans
Major Products:	Water Soluble Polymers
Wastewater Load:	COD - 40,000 lb/day Flow - 0.3 MGD
Celrobic System:	Start Up - Early 1977 Equalization - 0.2 MG Reactor - 0.3 MG Loading - 0.6 lb COD/ft^3/Day COD Removal Efficiency - 65% Methane Gas Value - $150,000/Yr ($3.5/MM BTU)

Table 2

Bishop Plant

Location:	Bishop, Texas (40mi SW Corpus Christi)
Major Feedstocks:	Natural Gas and Propylene
Major Products:	Methanol; Formaldehyde Acetate Esters; Butylene Glycol Pentaerythritol; Engineering Plastics
Wastewater Load:	COD - 100,000 lb/day Flow - 1 MGD
Celrobic System:	Start Up - Early 1981 Equalization - 1.5 Days Reactor - 1.5 MG Loading - 0.6 lb COD/ft^3/Day COD Removal Efficiency - 80% Methane Gas Value - $400,000/Yr ($3.5/MM BTU)

Table 3

Pampa Plant

Location:	Pampa, Texas (50mi NE Amarillo)
Major Feedstocks:	Butane Gas
Major Products:	Acetic Acid; Methyl Formate Acetic Anhydride; Ethyl Acetate Propionic Acid; Butyric Acid Methyl Ethyl Ketone; Formic Acid Methyl and Ethyl Acrylate
Wastewater Load:	COD - 120,000 lb/day Flow - 1 MGD
Celrobic System:	Start Up - Early 1981 Equalization - 1 Day Reactor - 1.5 MG Loading - 0.65 lb COD/ft^3/Day COD Removal Efficiency - 90% Methane Gas Value - $650,000/Yr ($3.5/MM BTU)

Table 4

Commercial Celrobic Installations

Company and Location	Date	Waste Source
Celanese Chemical Company Vernon, TX	1977	Guar Bean Processing
Celanese Chemical Company Bishop, TX	1981	Petrochemical
Celanese Chemical Company Pampa, TX	1981	Petrochemical
Nagata Sangyo Japan	1984	Wheat Starch
Enzyme Bio-Systems, Ltd. Beloit, WI	1985	Food-Fermentation
Hokuren, Japan	1985	Slaughter House
Shinko Shoji, Japan	1985	Piggery
Fujikko, Japan	1986	Food Processing
Nada, Japan	Under Construction	Food Processing

Figure 1

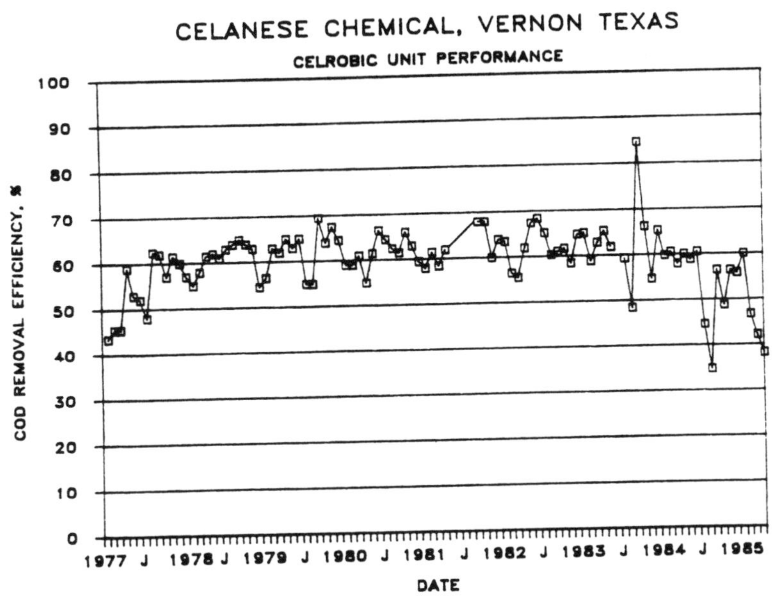

Figure 2

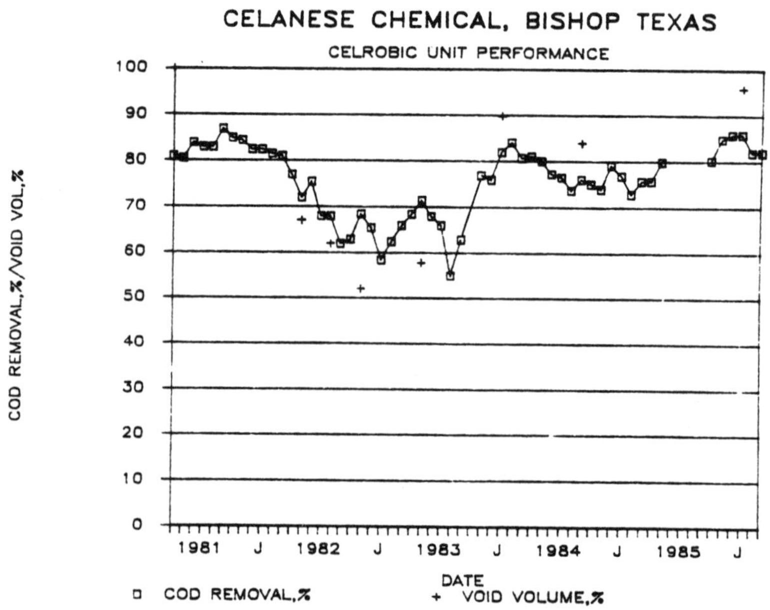

Figure 3

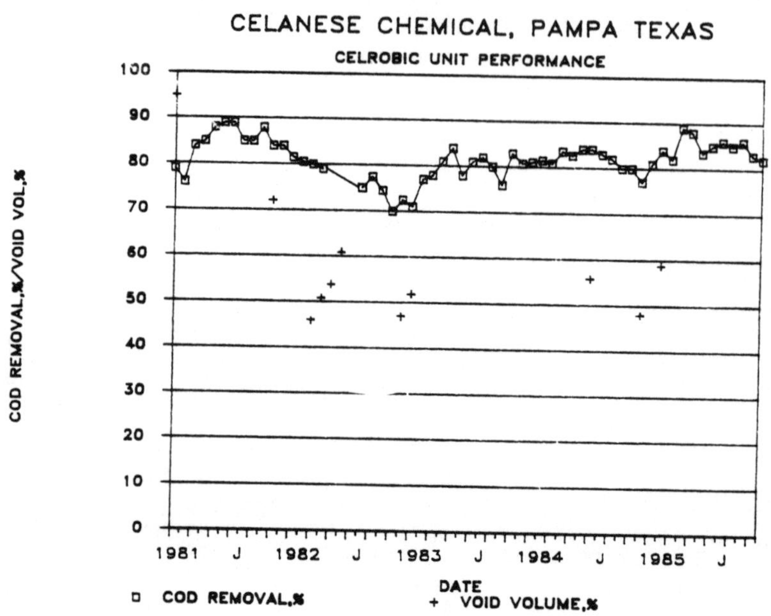

ANAEROBIC TREATMENT OF INDUSTRIAL WASTES

Adam S. Ng, Candace M. Rose, and Michael F. Torpy

Industrial Waste Research Laboratory
Energy and Environmental Systems Division
Argonne National Laboratory, Argonne, Illinois

ABSTRACT

Interest in anaerobic biotechnology for the treatment of industrial wastes has grown considerably. Anaerobic biological waste treatment offers advantages over aerobic systems in terms of lower energy requirements, less biological sludge production, and the potential for energy recovery in the form of methane gas. The development of innovative reactor designs, based on the optimization of growth and retention of anaerobic microorganisms, has created an impetus to reevaluate the anaerobic treatability of many industrial waste streams. The objective of the Industrial Waste Research Laboratory (IWRL) is to examine anaerobic treatability of industrial wastes, particularly those process-wastes originating from the Organic Chemical Production Industry. Data are presented to illustrate the potential applicability of anaerobic digestion for the treatment of a wide array of industrial process-waste streams.

TESTING PROGRAM

The testing program at IWRL consists of a series of laboratory and bench-scale tests designed to examine anaerobic treatability of industrial process-wastes. Laboratory Biochemical Methane Potential (BMP) Assays[1] and modifications of the conventional BMP procedure (to include waste-dependent pre- or co-treatment methods and/or the use of specifically adapted anaerobic inocula) are used for preliminary assessment of waste biodegradation potential. Subsequent bench-scale reactor testing allows for the determination of maximum rates of treatability, organic removal efficiencies, and process stability and requirements. Different bench-scale reactor designs, including batch-fed digesters, anaerobic contact processes, and high-rate fixed-film anaerobic processes, are used at the bench-scale testing level. The choice of a particular bench-scale design is determined by the physical nature of the process-waste sample under evaluation.

RESULTS

The chemical and physical characteristics of the process-waste samples tested were quite variable, resulting in varying levels of bioconversion efficiencies, ranging from 0 to 98 percent (see Table 1). In general, process-waste samples comprised of simple organics, such as alcohols, sugars, and organic acids, were readily amendable to anaerobic digestion. Some process-wastes required significant acclimation periods (i.e., 1-4 months) before the onset of bioconversion of organics. Other process-wastes, reported to contain certain materials (i.e., cyanide, acrylonitrile, heavy metals, etc.) that are known inhibitors of biological processes, were marginally or non-degradable. Chemical pretreatment methods to remove specific components in these samples or the use of specifically adapted microorganisms were generally unsuccessful in increasing digestibility.

At the bench-scale testing level, COD removal efficiencies exceeding 85 percent, at organic loadings of 10 g COD/L·day or greater, were observed for several Organic Chemical Production process-wastes (see Table 2). Some of these process-wastes required only simple neutralization and alkalinity supplementation to ensure high-rate treatability and process stability.

TABLE 1. Results of BMP Assay Testing

Sample Origin	Typical Components	COD (mg/L)	pH	Alkalinity (mg/L)	VFA (mg/L)	Total Solids (mg/L)	Assay Organic Concentration (COD, mg/L)	Bioconversion Efficiency (%)	Pre/co-Treatment Methods	Effect of Pre/co-Treatment Methods
Polyester production	Dimethylphthalate Terephthalate	3,000	3.97	140	570	1,056	500 1,000 1,500	71.0 74.0 38.0	— — —	— — —
Adipic acid production	Nitric acid, organic acids, Cu, V	4,300	1.8	—	—	5,192	500 1,000 1,500 2,000	52.0 72.0 55.0 62.5	— — — —	— — — —
Plasticizer/stabilizer production	Terpenes, organic acids, olefins	4,900	5.49	3,400	2,160	9,922	500 1,000 1,500	90.4 98.0 98.0	— — —	— — —
Acrylonitrile production	Thiocynates, inorganic-N, acrylonitriles, heavy metals, sulfates	2,800	3.47	Negligible	230	1,970	300 500 1,000 1,500	0 0 0 0	Sulfate removal by chemical ppt., sulfide precipitation with ferric chloride, CN removed by acid or oxidation, addition of PAC, use of different microbial strains	All pre/co-treatment methods were unsuccessful in increasing digestibility of the sample.
Polyethylene/fluoropolymer production	Chlorinated hydrocarbons, fluoropolymers	90,000	11.4	2,650	10	Negl.	500 1,000 1,500	35.1 67.0 12.5	Co-treatment with powdered activated carbon	No significant increase in digestibility.
Edible oil refining	Hydroxide & carbonate extracted raw oil	11,400	11.03	800	40	8,400	500 1,000 1,500	88.9 93.3 96.7	— — —	— — —
Polyethylene/fluoropolymer production	Fluoropolymers, methanol	9,000	3.7	50	10	Negl.	500 1,000 1,500	69.3 94.7 45.7	Co-treatment with powdered activated carbon	No significant increase in digestibility.
Organic chemical waste treatment solids	Coagulated-treated waste activated sludge	31,000	6.6	2,240	785	26,000	500 1,000 2,000 4,000	47.0 39.0 39.0 40.0	— — — —	— — — —
Organic chemicals production (raffinate stripper bottoms)	Acrylic acid, acetic acid, chromium	25,250	2.1	—	—	Negl.	500 1,000 1,500	0 0 0	EDTA-treated to chelate chromium —	No increase in sample digestibility

TABLE 1. Results of BMP Assay Testing (cont'd)

Sample Origin	Typical Components	COD (mg/L)	pH	Alkalinity (mg/L)	VFA (mg/L)	Total Solids (mg/L)	Assay Organic Concentration (COD, mg/L)	Bioconversion Efficiency (%)	Pre/co-Treatment Methods	Effect of Pre/co-Treatment Methods
Batch specialty organic chemical production	Isopropyl alcohol, naptha, toluene, xylene, methanol	215,500	9.4	9,400	6,100	Negl.	500 1,000 2,000	42.7 18.0 11.0	Co-treatment w/PAC, use of innocula acclimated to industrial wastes, petroleum sludges, xylene and toluene	The addition of PAC increased digestability of sample (1,000 g COD/L) from 18% to 81.3%.
Organic chemicals production process-waste	Acetic acids, ammonia, high acidity, chromium	20,250	3.76	—	2,630	Negl.	500 1,000 1,500	0 0 6.1	EDTA-treated to chelate chromium	No increase in sample digestibility.
Organic chemical waste treatment solids	Coagulated-treated filter-pressed biological solids	0.875 g COD/ g solids	7.07	920	—	12%	500 1,000 1,500	33.0 36.0 37.0	— — —	— — —
Chemical production process-waters	Unknown (high acidity waste)	88,000	0.3	—	—	136,000 (white ppt)	500 1,000 1,500	0 0 0	Sample separation into two settlable layers; use of "sludge thickener" seed innoculum	Sample digestibility was increased slightly 8% degraded) with the use of the sludge thickener seed.
Agricultural chemical production wastes	Alachlor, butachlor, di-methylaniline, formaldehyde methanol, butanol	3,000	5.56	410	130	Negl.	500 1,000 1,500	10.4 25.9 5.4	Use of seed innocula acclimated to the presence of 50 mg/L formaldehyde	Increased sample digestibility to 21% at the lowest concentration tested (500 mg COD/L).
Agricultural chemical production waste	Alachlor, dimethylamine, formaldehyde, methanol	6,500	6.8	420	10	Negl.	500 1,000 1,500	0 12.5 10.7	Use of seed innocula acclimated to the presence of 50 mg/L formaldehyde	No significant increase in sample digestibility, no CH_4 formed.
Adiponitrile production	Organic nitriles, Cn, Nickel, H_2SO_4, Boron	4,400	9.75	5,640	130	1,665	500 1,000 1,500	Ongoing	S_2O_3 & Fe addition; CN oxidation; use of "sludge thickener" seed from chemical processing plant.	Ongoing

CONCLUDING REMARKS

Several industrial process-wastes were found to be highly amendable to anaerobic digestion. Some process-wastes required only simple pretreatment (i.e., neutralization, alkalinity supplementation, nutrient additions) to ensure process-stability. Other process-wastes were found to be unsuitable for anaerobic treatment because of the presence of toxic or inhibitory materials. Important factors which may determine the potential for high-rate treatability include the choice of reactor/process configuration, pre- and/or co-treatment conditions, nutrient supplementation, and microbial acclimation ability.

ACKNOWLEDGMENTS

This work was supported by the U.S. Department of Energy, Assistant Secretary for Conservation and Renewable Energy, Industrial Waste Products Utilization Programs, under Contract W-31-109-Eng-38.

REFERENCE

[1] Owen, W.F., Stuckey, D.C., Healy, J.B., Young, L.Y., and McCarty, P.L., "Bioassay for Monitoring Biochemical Methane Potential and Anaerobic Toxicity", Water Research, Vol. 13, 485-492, 1979.

Table 2: COD Removal Efficiencies

Sample Origin	Typical Components	COD (mg/L)	Reactor Type	Max. Organic Loading Rate Tested	% COD Removal	Comments on Treatability
Polyester Production	Dimethylphthalate teraphthalate	2350	Upflow anaerobic filter	4.8 g COD/L.Day (low sample COD limited testing of higher OLRs)	60-78	Microbial acclimation important for high-rate treatment; supplemental nutrients, alkalinity requirements
Plasticizer/ Stabilizers production	Acetic acid, formic acid, aldehydes, 1% sodium sulfate, 0.05% ferrous-ferric sulfate/hydroxide	4900	Upflow and downflow anaerobic filter	20 g COD/L.day	83-94	Optimal OLR tested was 15.0 g COD/L.day with 94% COD removals.
Ethanol production facility	Polysaccharides, ethanol	22500	Anaerobic-contact process	3.4 g COD/L.day	90	>2% total solids; operation at higher OLR is probable with different reactor process.
Organic Chemical waste treatment solids	coagulant-treated biological solids	31000	Anaerobic-contact process	4.4 g VS/L.day	Max VS destruction of 1.5 g VS/L.day at OLR OF 3 g VS/L.day	2.8% total solids; inorganic scum layer formation.
Edible oil refining	Hydroxide & carbonate extracted oils	12000	Upflow anaerobic filter	12.5 g COD/L.day	85.3% @ OLR of 10 g COD/L.day	Alkalinity supplementation required for process stability.
Organic chemical waste treatment solids	Coagulant-treated fliter-pressed biological solids	0.88 g COD/ g dry solids	Batch-fed anaerobic digester	0.55 g COD/L.day	<10	Coagulant-treatment reduces sample digestibility
Adipic acid production	nitric acids, organic acids, Cu, V	4900	Downflow anaerobic filter	4.6 g COD/L.day (ongoing)		Expect sequential denitrification/ methanogenesis

Other Noyes Publications

MINIMIZING EMPLOYEE EXPOSURE TO TOXIC CHEMICAL RELEASES

by

Ralph W. Plummer, Terrence J. Stobbe, James E. Mogensen, Luanne K. Jeram
West Virginia University

Pollution Technology Review No. 145

This book describes procedures for minimizing employee exposure to toxic chemical releases and suggested personal protective equipment (PPE) to be used in the event of such chemical release. How individuals, employees, supervisors, or companies perceive the risks of chemical exposure (risk meaning both probability of exposure and effect of exposure) determines to a great extent what precautions are taken to avoid risk.

The objective of this study was to obtain information from chemical manufacturers and facilities which use chemicals in other processes, to determine what types of procedures are currently in practice, and to develop a set of recommended procedures and suggested PPE which can be used to minimize the possibility and the risk of toxic chemical releases. In Part I, the authors develop an approach which divides the project into three phases: kinds of procedures currently being used; the types of toxic chemical release accidents and injuries that occur; and, finally, integration of this information into a set of recommended procedures which should decrease the likelihood of a toxic chemical release and, if one does occur, will minimize the exposure and its severity to employees. Part II covers the use of personal protective equipment. It addresses the questions: what personal protective equipment ensembles *are used* in industry in situations where the release of a toxic or dangerous chemical may occur or has occurred; and what personal protective equipment ensembles *should be used* in these situations.

A condensed table of contents listing **part titles, chapter titles and selected subtitles** is given below.

ISBN 0-8155-1131-0 (1987) 257 pages